Nanomaterials for Applications in Energy and Biology 2017

Topical Conference at the 2017 AIChE Annual Meeting

Minneapolis, Minnesota, USA
29 October - 3 November 2017

ISBN: 978-1-5108-5783-4

Printed from e-media with permission by:

Curran Associates, Inc.
57 Morehouse Lane
Red Hook, NY 12571

Some format issues inherent in the e-media version may also appear in this print version.

Printed by Curran Associates, Inc. (2018)

AIChE
120 Wall Street, FL 23
New York, NY 10005-4020

Phone: (800) 242-4363
Fax: (203) 775-5177

www.aiche.org

Additional copies of this publication are available from:

Curran Associates, Inc.
57 Morehouse Lane
Red Hook, NY 12571 USA
Phone: 845-758-0400
Fax: 845-758-2633
Email: curran@proceedings.com
Web: www.proceedings.com

TABLE OF CONTENTS

Some abstracts in this book refer to images that were not made available by the publisher

493137 Self-Assembling Nanoparticles for Peptide Delivery with Enhanced Stability

Abstract:

Peptides are promising tools for manipulating disease processes within cells, but their delivery to and into cells remains a major obstacle. Peptide amphiphiles (PAs), a peptide conjugated to a hydrophobic tail, are one tool to enable the intracellular delivery of therapeutic peptides. The hydrophobic tails drive nanoparticle (NP) self-assembly, which are optimally sized (10-20 nm) to promote accumulation in tumor tissues via the enhanced permeability and retention effect, and their structure prevents hydrolysis and proteolysis of peptides in circulation. The internalization mechanisms have not been fully elucidated, but it is known that endocytosis plays an important role. Despite these delivery benefits, internalized PA NPs mostly remain trapped in endosomes preventing them from reaching their cytoplasmic targets and thus reducing their therapeutic efficacy. Introduction of endosomal cleavage and escape groups into the building blocks of the nanoparticles requires intake through endocytosis.

An additional challenge to overcome for the PA NPs is their limited circulation time in the body. NP formation of PAs occurs through the hydrophobic interactions of the tail. These interactions are strong enough to form NPs in aqueous medium. Nevertheless, the hydrophobic components in body serum cause the disassembly of the NPs. Because of this reason, *in vitro* experiments should be performed in FBS free medium. Free PAs are vulnerable to proteases and too small to internalize through endocytosis. As a result, the stability of the NPs should be improved to ensure the increased circulation period and endocytosis.

Therefore, we studied the stability of engineered self-assembling nanoparticles with controlled disassembly characteristics. We mixed biocompatible UV cross-linkable hydrophobic components with peptide amphiphiles prepared with DSPE-PEG(2000) during the self-assembly process. The crosslinking components were encapsulated in the core and enhanced the stability through entanglement of the hydrophobic tails after UV crosslinking. We controlled the stability of the self-assembling NPs against temperature by altering the crosslinking components. Furthermore, we showed the enhanced circulation time by *in vivo* experiments. This design enhances the stability of NPs after self-assembly through crosslinking the core. Thus, the engineered PA located on the NP surface is not affected by the crosslinking. Furthermore, this strategy can enable the intake of NPs through endocytosis, which is convenient for engineering endosomal cleavable and escape domains to deliver the therapeutic peptides.

494457 Block Copolymer Nanoparticle for Biofilm Dispersal of Gram-Positive MRSA and VRE

Abstract:

Biofilm-associated infections have become a major threat to human health and bacteria cells inside biofilms are typically around 1000-fold more resistant to antibiotics. Therefore, the conventional antibiotics or antimicrobial peptides(AMPs) become less effective to work on the biofilm bacteria. It is our priority to develop new antibiofilm agents with good biocompatibility. In order to address this problem, a novel anti-biofilm cationic polymeric nanoparticles have been studied in our group, and these nanoparticles showed excellent antibiofilm activity and are hemolytic with good therapeutic index as well as good *in vivo* cytotoxicity. Dextran and poly(3-Acrylamidopropyl) trimethylammonium chloride (AMPTMA)) are both hydrophilic polymers and we attempted to explore the anti-biofilm effects of copolymers made from these, and with a minor component based on poly (butyl methacrylate) (PBMA). Four types of AMPTMA-based copolymers were made by the facile Atom Transfer Radical Polymerization (ATRP) reaction: (i) A100 which is poly(AMPTMA)100; (ii) DA100 which is a dextran-block-poly (AMPTMA)100; (iii) DAB which is dextran-block-poly (AMPTMA-co-BMA) and (iv)AB which is poly(AMPTMA-co-BMA). For DAB, it can self-assemble into nanoparticles which are 20-30 nm in diameter with the dextran block as the corona, and the poly poly(AMPTMA-co-BMA)) as core. However, the rest three polymers didnâ€™t form this nanoparticle. Our results showed that DA95B5 copolymers with micellar nanostructure are effective in the dispersal of preformed biofilms of clinically relevant Gram-positive *MRSA* and *Vancomycin-Resistant Enterococci (VRE V583)* with low *in vivo* cytotoxicity. To the best of our knowledge, this is the first report of cationic polymeric micellar nanoparticle as good antibiofilm agents with effective removal of pre-formed biofilm with good biocompatibility in contrast to many other such agents which are enzyme-based.

495726 Genetically-Targeted Brain-Machine Interface

Abstract:

Interfacing with living neural circuits using high-speed electrophysiological tools has contributed substantially to basic neuroscience as well as to clinical neurology and psychiatry. However, the limited ability of these devices to interface with the intact living brain with specificity for cell types prevents both deeper understanding of neural circuit function and much needed precision for clinical intervention. Here, we present an unprecedented approach for a direct formation of electrical connections with targeted cells through the convergence of genetic engineering and polymer chemistry. With this method, novel genetically-targeted electronics are intrinsically assembled at the surface membrane of neurons both within functional neural tissue and in human 3D brain cultures derived from stem cells. This approach advances the field toward cell type-resolution electrical tuning of local neuronal activity, via bridging brain regions to external recording devices by precisely-defined high-speed pathways. Moreover, by enabling assessment and control of neural activity via fast genetically-targeted electrical connections, this technology may help broadly advance the understanding of neural circuit operation in health and disease, new treatment strategies for neuropsychiatric disease, and acceleration of the path toward specifically-enhanced sensation, cognition, and performance.

498220 Colorimetric Detection of Nitrite Ions Based on the Aggregation of Gold Nanoparticles

Abstract:

Many bacteria associated with urinary tract infections (UTIs) produce enzymes that reduce nitrate to nitrite. After the presentation of clinical symptoms, UTI diagnosis is based on a point-of-care (POC) test, the urine dipstick. Nitrite detection is a key component of these POC tests. However, interference, variable capacity of bacteria to reduce nitrites, diet, or dwell time of urine in the bladder likely lead to lower reliability and accuracy of nitrite testing with diagnostic sensitivity being reported between 10 and 93%. Although urine cultures or advanced characterization methods, such as chemiluminescence, electrochemistry, and surface-enhanced Raman scattering, are more definitive in diagnosis, they are slow tests that could lead to the deterioration of patients from delayed treatment. For this reason, POC testing remains an important diagnostic step in UTI management. Here we describe new, simple and highly sensitive colorimetric assays that provide visual on-site analysis. These facile colorimetric assays are based on the anti-aggregation of gold nanoparticles. The gold nanoparticles are first coated with functional ligands that cause an immediate color change to blue, which is associated with widespread aggregation of the nanoparticles. After the addition of nitrite ions, the nanoparticles are re-dispersed in solution, yielding a bright red or purple color that is similar to the initial nanoparticle suspension. By varying the concentration of the ligands and nanoparticle size, the sensitivity of these nitrite sensors is greatly increased, achieving order of magnitude improvements in detection limits. Furthermore, the effect of this assay on synthetic urine has been explored to understand the effect of foreign ions on the detection of nitrite.

498301 Cancer Cell Hyperactivity and Membrane Dipolarity Monitoring Via Raman Mapping of Interfaced Graphene: Towards Non-Invasive Cancer Diagnostics

Abstract:

Ultra-sensitive detection, mapping and monitoring of the activity of cancer cells is critical for treatment evaluation and patient care. Here, we demonstrate that a cancer cell's glycolysis-induced hyperactivity and enhanced electronegative membrane (from sialic acid) can sensitively modify the second-order overtone of in-plane phonon vibration energies (2D) of interfaced graphene *via* a hole-doping mechanism. By leveraging ultrathin graphene's high quantum capacitance and responsive phononics, we sensitively differentiated the activity of interfaced Glioblastoma Multiforme (GBM) cells, a malignant brain tumor, from that of human astrocytes at a single-cell resolution. GBM cell's high surface electronegativity (potential ~310 mV) and hyperacidic-release induces hole-doping in graphene with a 3-fold higher 2D vibration energy shift of approximately 6 ± 0.5 cm^{-1} than astrocytes. From molecular dipole induced quantum coupling, we estimate that the sialic acid density on the cell membrane increases from one molecule per ~17 nm^2 to one molecule per ~7 nm^2. Further, graphene phononic response also identified enhanced acidity of cancer cell's growth medium. Graphene's phonon-sensitive platform to determine interfaced cell's activity/chemistry will potentially open avenues for studying activity of other cancer cell types, including metastatic tumors and characterizing different grades of their malignancy.

489494 Functionalization of Graphene By Using Protein Engineering

Abstract:

We have designed peptide specific in binding to the graphene surface non-covalently. We believe this would be helpful in designing sensors that could be advantageous in detecting specific molecules of interest. In this study, we employed surface characterization of graphene surface using time of flight secondary ion mass spectroscopy, atomic Force Microscopy and theoretical biophysics studies to investigate the peptide-graphene interface interactions. We found peptide forms self-assembly on graphene surface, and have high specificity towards graphene surface forming hydrophobic and Ï€-Ï€ interactions. Further, we introduced another protein on the functionalized peptide-graphene surface to check the sensing ability of the device, our results shows the device could be advantageous in designing accurate and specific biosensors for comprehensive analysis of protein molecules.

496391 Developing Precision Medicine Using Quantum Biology: Combining Quantum States, Surface Chemistry, and Microbiology

Abstract:

Health and medicine are personal for each individual, but most of the current diagnosis and treatments follow "one size fits all" approach. Recently there has been increasing awareness to remedy this, and several initiatives have been started towards personalized medicine, precision medicine, and others, to develop customized and tailored solutions. In this talk, I will address recent advances made in my group in developing precision medicine using "Quantum Biology", a set of spectroscopic techniques and quantum-confined materials addressing molecular and nanoscale material and surface interactions to develop transformative solutions for detection, diagnosis and treatment of diseases. Particularly, my talk will focus on developing a new class of nanomaterial therapeutics for drug-resistant superbugs that selectively target the pathogens, without adversely affecting the growth of nearby human host cells. These therapeutics can be triggered by an external stimuli, which can allow selective targeting and controlled drug delivery, while reducing possible adverse side-effects.

Multidrug-resistant bacterial infections are an ever-growing threat because of the shrinking arsenal of efficacious antibiotics, and the high-frequency of multidrug-resistant (MDR) bacterial infections combined with the lack of new antibiotics threatens the future of our healthcare system as we approach a post-antibiotic era. An increasing class of gram-negative multidrug-resistant pathogens like Enterobacteriaceae including carbapenem-resistant (CRE) *Escherichia coli* and extended spectrum β-lactamase (ESBL) producing *Klebsiella pneumoniae* (KPN) are severely antibiotic resistant and were recently designated priority 1 critical class bacterial pathogens in urgent need of effective antibiotics by the World Health Organization. To address an urgent need for developing a new class of antibiotics beyond the nominal small molecule discovery, and adjuvants to work or potentiate existing antibiotics, my group is developing nanomaterial therapeutics combining quantum confined materials, surface chemistry, and microbiology. Furthermore, while using external stimulus like light has been successful in addressing issues of drug delivery and transport in metal nanoparticle based therapies, metal nanoparticles induced cell death and toxicity effect is typically nonspecific. Here, I will show a range of quantum confined nanomaterials such as quantum dots (QDs) can kill a wide range of multidrug-resistant bacterial clinical isolates, including methicillin-resistant *Staphylococcus aureus*, carbapenem-resistant *Escherichia coli*, and extended-spectrum β-lactamase-producing *Klebsiella pneumoniae* and *Salmonella typhimurium*. The killing effect is independent of material and controlled by the redox potentials of the photogenerated charge carriers, which selectively alter the cellular redox state. We also show that the QDs can be tailored to kill 92% of bacterial cells in a monoculture, and in a co-culture of *E. coli* and HEK 293T cells, while leaving the mammalian cells intact, or to increase bacterial proliferation. These studies can not only lead to development of new antibiotics and adjuvants for clinical therapeutics for the treatment of infections, but also be used in the study of the effect of redox states on living systems.

NANOSCALE SIZE-CONTROLLED ELECTROSPUN COMPOSITE MATS OF CHITOSAN WITH TUNGSTEN DISULFIDE INORGANIC NANOTUBES (INT-WS_2)

A. Baklavaridis[1,3], I. Zuburtikudis[2,3,], C. Panayiotou[1]*

[1]Department of Chemical Engineering, Aristotle University of Thessaloniki, Greece

[2]Department of Chemical Engineering, Abu Dhabi University, Abu Dhabi, U.A.E.

[3] Dept. of Mechanical and Industrial Design Engineering, Western Macedonia University of Applied Sciences, Kozani, Greece

Abstract

In this work, we report the tailor-made fabrication of Chitosan nanofibrous mats reinforced with tungsten disulfide inorganic nanotubes (INT-WS_2). The objective was to research the effect of the electrospinning process and material parameters on the obtained morphology of the mats. The INT-WS_2 content, the polymer solution concentration, the electric field strength and the solution's flow rate were the factors investigated within the framework of Response Surface Methodology (RSM). SEM and image analysis were employed for the dimensional characterization of the nanofibrous morphology and the estimation of three selected responses. Two responses were related to the quality of the nanofibrous morphology: the number surface density of the beads (N_{bead}) and the average bead-to-fiber diameter (D_{bead}/D_{fiber}). The third response was the fiber thickness (D_{fiber}). The results indicate that the polymer solution concentration, the electric field strength and the INT-WS_2 loading are the factors that mostly affect the electrospun nanofibrous morphology. The response-surface results revealed possible experimental pathways that may be followed in order to obtain a desired nanofibrous chitosan/INT-WS_2 morphology for the intended biomedical application.

1. Introduction

Electrospinning is an attractive technique for the production of highly porous and fibrous structures [1,2]. These fibrous materials have been successfully used in membrane/filtration technology [3,4] and biomedical applications [5,6]. Moreover, electrospinning, compared to other techniques, has many advantages: high production yield, low-cost and the simplicity of the process [7,8]. The fiber diameters that can be produced by electrospinning may vary from a few nanometres to a few micrometres.

Nanofibers cover a broad range of current and emerging technological applications which include tissue engineering, anti-microbiological wound dressings, adsorption/filtration, drug delivery [9-11], catalysis [12], and sensors [13]. In many of the aforementioned applications, the biodegradability and the biocompatibility of the polymer should be considered. Chitosan is a biodegradable and biocompatible carbohydrate polymer with the ability to form nanofibers via the electrospinning technique [14-16]. Electrospinnable chitosan solutions have been prepared using various solvents such as aqueous acetic acid [15-19], trifluoroacetic acid [20], and trifluoroacetic acid/dichloromethane mixtures [20-24]. Aqueous acetic acid should be the preferred solvent in the case of large-scale chitosan nanofiber production due to its low cost and due to reduced environmental and health considerations compared to the rest of the aforementioned solvents [4]. However, difficulties in obtaining uniform chitosan nanofibers from aqueous acetic acid solutions have been reported [4, 20, 25].

---

* Corresponding Author: Ioannis Zuburtikudis, E-mail: ioannis.zuburtikudis@adu.ac.ae & ioanzub@gmail.com

Nanocomposite nanofibers have attracted much attention lately, due to the significant enhancement in the mechanical and other properties of nanofibers [27-30]. Several nano-fillers can be used as reinforcements for the production of nanocomposite nanofibers depending on the targeted application. In the case of chitosan, multiwall carbon nanotubes (MWCNTs) [23, 31], hydroxyapatite [32, 33], mesoporous silica [34], silver nanoparticles [25] and chitin nanocrystals [35] have been used. Therefore, the continuation of the research efforts aiming at the nano-reinforcement of chitosan nanofibers may expand their use to new applications and products. However, studies dealing with chitosan nanocomposites that incorporate other types of nano-fillers and especially nanotubes (e.g. inorganic nanotubes) are still missing. Tungsten disulfide inorganic nanotubes (INT-WS_2) is an interesting type of nanotubes, which combine a variety of attractive physiochemical properties such as biocompatibility, high Young's modulus, compressive strength and chemical inertness [36-38]. At the same time, the nanotubes' effect on the electrospun morphology was not thoroughly investigated.

For the effective development and commercialization of the electrospun mats, in order to be used in the aforementioned technological applications, someone has to consider not only the fibrous structure itself but also the beads which often occur in the fibrous structure. Even though beads are regarded as non-desirable characteristics ("byproducts") of the fibrous structure, they can have a functional role in various applications. It has been reported that beads on the electrospun fibrous structure can be used for cell inclusion or drug reservoirs, while the beads exhibit different degradation behavior than the neat fibers. Consequently, apart from the fiber diameter examination and quantification, the density of the beads on the surface of the electrospun mat should be included to the analysis.

Many studies regarding the electrospinning of chitosan and other biopolymers [19, 20, 25, 39] investigate the effect of the electrospinning parameters on the fiber diameter by modifying one factor at a time, while maintaining the other factors constant. Although this methodology is rather straightforward, it cannot reveal potential interactions between the factors studied. Therefore, the applicability of design of experiments (DoE) statistical methodology in the study of the electrospinning process of a particular system is very attractive.

In this work, a novel polymer-nanotube nanocomposite system in the form of nanofibrous membrane was investigated. Chitosan nanofibrous membranes, reinforced with tungsten disulfide inorganic nanotubes (INT-WS_2), were prepared by electrospinning using DoE. The experimental design was employed for the following purposes: a) the identification and quantification of process and material parameters that affect significantly the final electrospun chitosan/INT-WS_2 nanofibrous morphology and their interactions, and b) revealing possible pathways in order to obtain nanocomposite and nanofibrous materials based on Chitosan/INT-WS_2 with the desired morphology.

2. Experimental Section

2.1. Materials.

A medical grade chitosan (95/500) with high deacetylation degree (≥92.6%) was purchased from Heppe Medical Chitosan GmbH (Germany). Acetic acid (100%, anhydrous) was purchased from Merck (Germany), while INT-WS_2 were purchased form Nanomaterials Ltd. (Israel). All materials were used as received, without any further treatment.

2.2. Preparation of the nanocomposite electrospun mats.

Initially, chitosan/INT-WS_2 solutions, needed for the electrospinning process, were prepared. Specified INT-WS_2 content (0.5-1% w/v) was dispersed in aqueous acetic acid (90% acetic acid/10% water v/v) by sonicating the solution for 10 minutes. Relatively low loadings of chitosan (2.5-3.5% w/v) were then added in the solution and subsequently sonicated for another 5 minutes. In the case of chitosan-only solution preparation the first step of the procedure (sonication for 10 minutes) was omitted. The (nanocomposite) chitosan solution was stirred overnight until homogenous solution was received. This homogeneous solution was used in the custom electrospinning set-up shown in Fig. 1. The setup was enclosed inside a chamber, so as to maintain a stable environment during the electrospinning process. The experiments were carried out at room temperature (23.4-25.1 °C) and ambient relative humidity (30-35%). The ambient conditions were monitored and recorded for every single experiment. The precise ambient temperature setting was ensured by using an air-conditioning unit in the laboratory, while a temperature and humidity sensor (attached inside the electrospinning chamber) was used in order to monitor the environment during the electrospinning process.

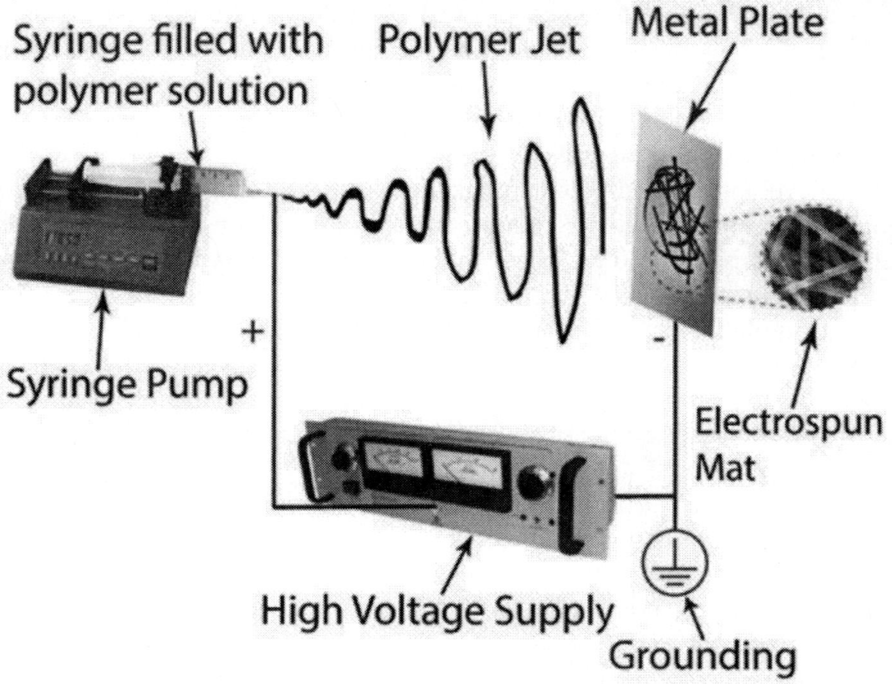

Figure 1: A schematic representation of the electrospinning used.

2.3. Experimental Design and Responses.

A D-optimal experimental design was used, with an initial condition number equal to 8.37 (<10) and objective the response-surface modeling. This experimental design included 29 experimental runs, while four of these runs were pure replicates. The following four factors were selected as controllable factors: the flow rate (qualitative factor), the solution concentration (quantitative factor), the INT-WS_2 content in the polymer solution (quantitative factor) and the electric field strength (voltage/length unit - quantitative factor). In this case, the flow rate was considered as a qualitative factor, due to the fact that the experimental setup used allowed the flow rate to be defined only at specified levels.

The whole experimental workflow and analysis are described in Fig. 2 and details can be found in ref. [40].

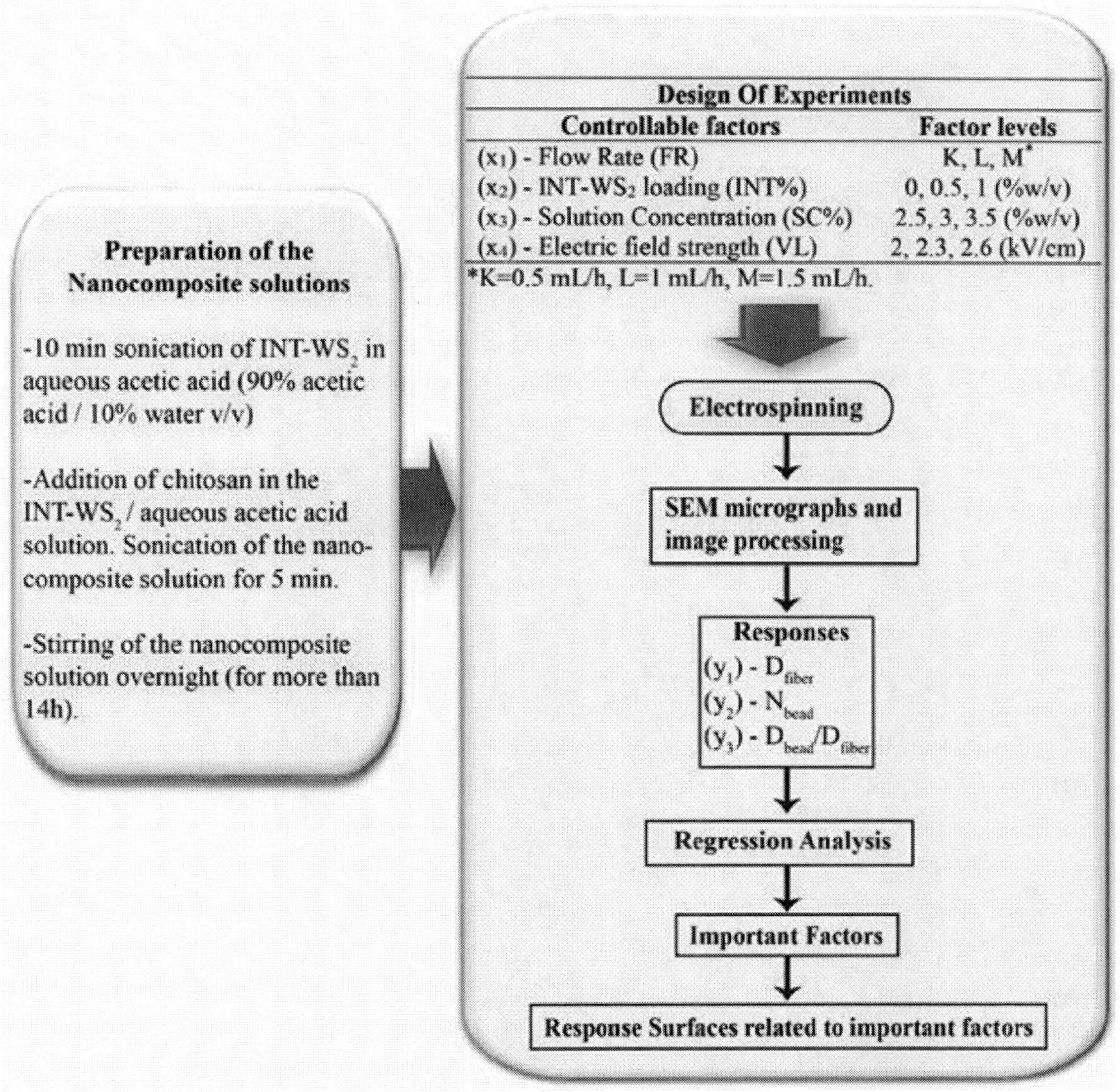

Figure 2. An illustration of the experimental workflow, in the framework of the design of experiments, and the analysis of the results carried out.

Following the procedures described in fig. 2, the electrospun mats from the 29 experimental runs were fabricated and their SEM images were taken. The images were processed in order to quantify their morphology (responses: mean fiber diameter D_{fiber}, number density of beads N_{bead}, and the ratio D_{bead}/D_{fiber}). Subsequently, the responses were used for the regression analysis, which resulted in the identification of the important factors. Thus, the contour plots (response surfaces) of the important factors were made. More details for the experimental procedures followed, the regression analysis and the model developed, and the contour plots derived can be found in ref. [40].

3. Results

Fig. 3 presents SEM images, which show the different morphologies produced. The quantification of the morphology of the mats and the regression analysis revealed that the solution concentration (SC%), the electric field strength (VL) and the INT-WS_2 loading (INT%) significantly affect the electrospun nanofibrous morphology. The contour plots for all responses as a function of %SC and %INT at the L level of FR and at three values of VL (2, 2.3, 2.6 V/cm) are presented in Fig. 4.

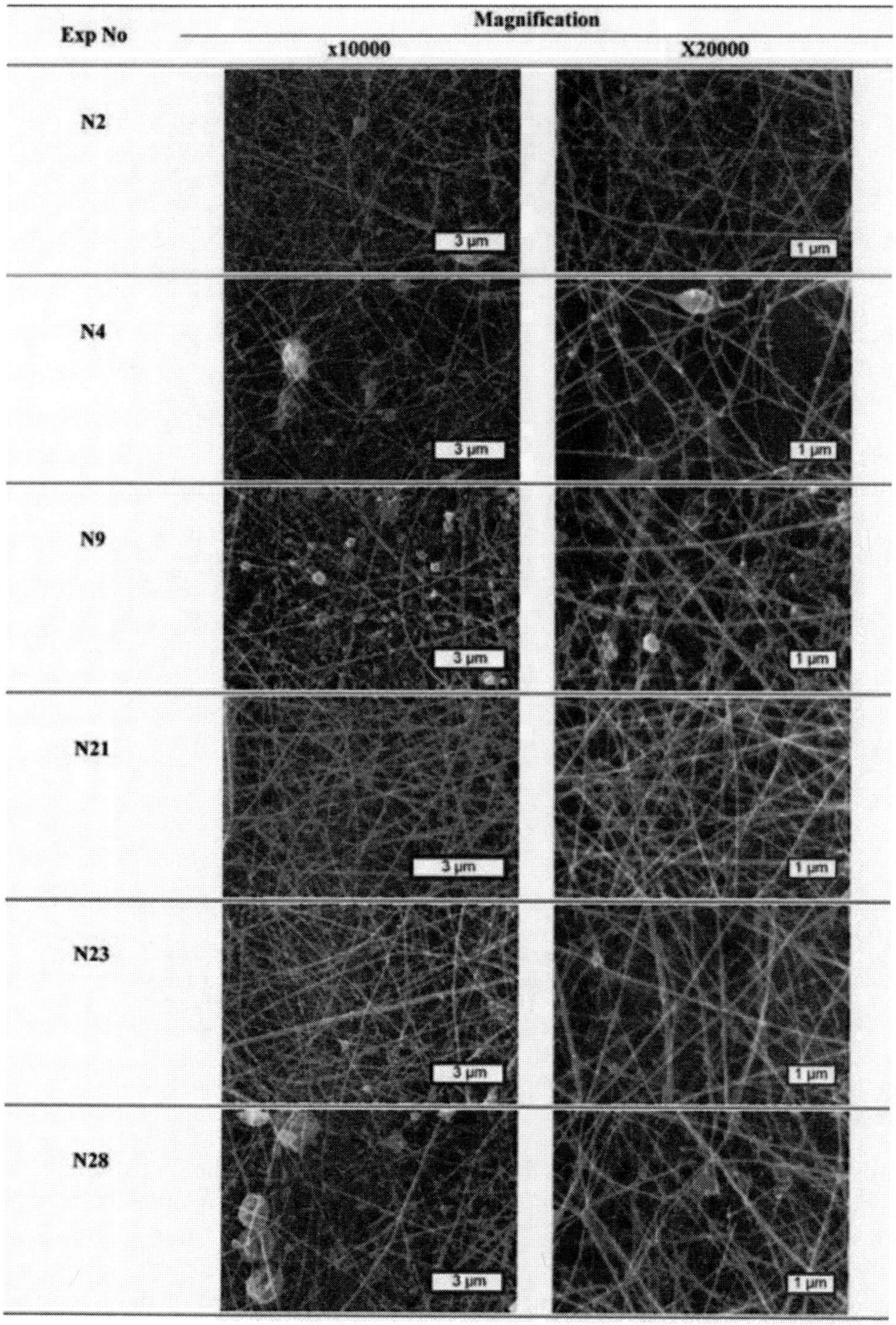

Figure 3. SEM images of 6 morphologically indicative experimental runs (N2, N4, N9, N21, N23, N28). The scale bar for the x10000 images is set to 3 μm, while for the x20000 is set to 1 μm.

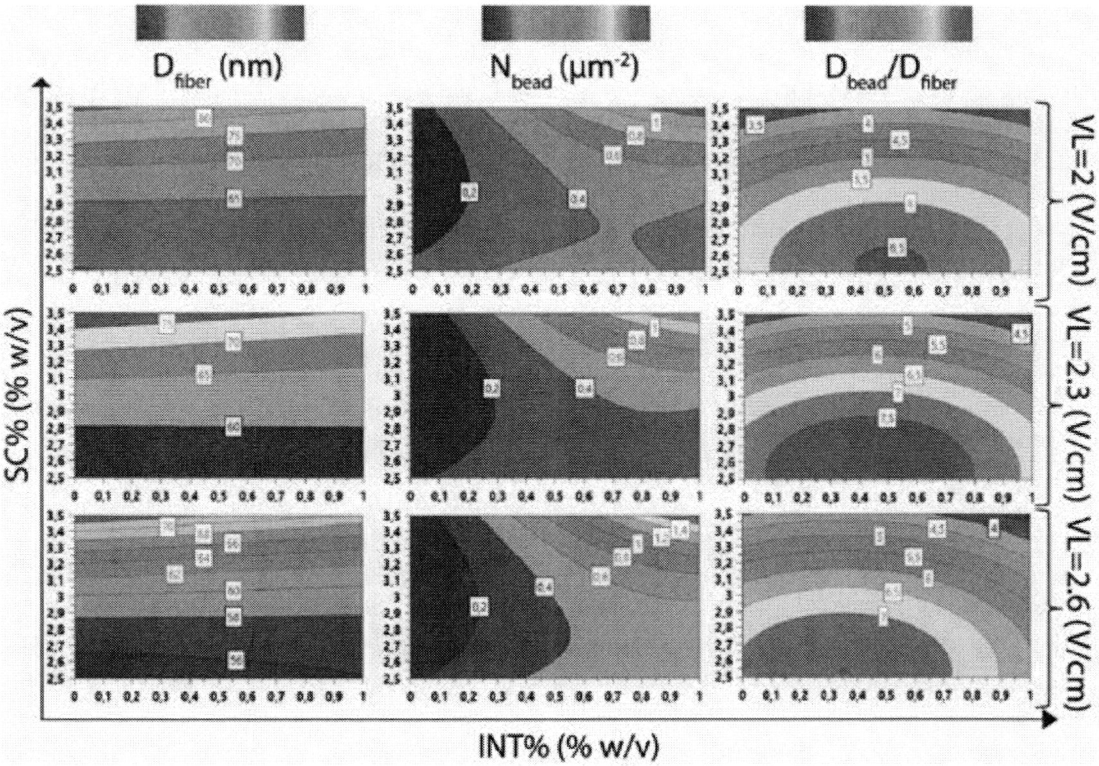

Figure 4. Contour plots for all responses as a function of %SC and %INT at the L level (1 mL/h) of FR and at three values of VL (2, 2.3, 2.6 V/cm). Specifically, each column of the figure illustrates contour plots that present the variation of a response (D_{fiber}, N_{bead}, or D_{bead}/D_{fiber}) as a function of the two aforementioned factors (%SC and %INT). Each row depicts contour plots for the three responses that correspond to a specific value of VL (2, 2.3 or 2.6 V/cm). The different colors in the plots denote the response value's fluctuation from low (blue) to high (red) values.

For the contour plots of the three responses for electric field strength (VL) 2 V/cm, nanofibers in the range 56-65 nm may be produced at low polymer solution concentrations. At high solution concentrations the fiber diameter (D_{fiber}) increases to 80 nm. Nevertheless, the INT-WS_2 loading does not affect D_{fiber} due to the fact that the contours are situated collaterally among each other and also situated collaterally with regard to the horizontal axis (INT%). On the other hand, the bead density (N_{bead}) is rather low (0.2 beads/μm) in very low INT-WS_2 loadings, while in high INT-WS_2 loadings N_{bead} becomes quite high (1 beads/μm). Adding to that, the combined effect of the INT-WS_2 loading and the polymer solution concentration, previously mentioned, here becomes evident: N_{bead} becomes quite high (1 beads/μm) if only the polymer solution concentration is also rather high (~3.5% w/v). Moreover, at very low polymer solution concentrations (2.5% w/v) the N_{bead} is almost double than the N_{bead} for intermediate polymer solution concentrations (3% w/v) and for a certain INT-WS_2 loading. This means that if someone wants to produce rather bead-free INT-WS_2 nanocomposite chitosan nanofibers, a polymer solution concentration around 3% w/v and no or very low INT-WS_2 loadings should be chosen. This is also in accordance with our experimental findings/morphological observations. The contour plot which corresponds to the bead-to-fiber diameter (D_{bead}/D_{fiber}) has contours which are ellipsoidally formed. Hence, D_{bead}/D_{fiber} values are increasing when the solution concentration is low and the INT-WS_2 loading obtains intermediate values (0.5 % w/v). The D_{bead}/D_{fiber} values are decreasing at higher polymer solution concentrations and the lowest possible D_{bead}/D_{fiber} values are recorded at the corners of the graph (at very low and at very high INT-WS_2 loadings). The response-surface results revealed possible experimental pathways that may be followed in order to obtain a desired nanofibrous chitosan/INT-WS_2 morphology for the intended biomedical application.

References

1. A. Greiner and J. H. Wendorff (2007), *Angewandte Chemie - International Edition*, **46**, 5670.
2. D. H. Reneker and I. Chun (1996), *Nanotechnology*, **7**, 216.
3. N. N. Bui, M. L. Lind, E. M. V. Hoek, and J. R. McCutcheon (2011), *Journal of Membrane Science*, **385-386**, 10.
4. L. Li, Y. Li, L. Cao, and C. Yang (2014), *Carbohydrate Polymers*.
5. M. Sadri, S. Arab-Sorkhi, H. Vatani, and A. Bagheri-Pebdeni (2015), *Fibers and Polymers*, **16**, 1742.
6. L. Bosworth and S. Downes (2011), "Electrospinning for Tissue Regeneration", Elsevier Science.
7. J. H. Wendorff, S. Agarwal, and A. Greiner (2012), "Electrospinning: Materials, Processing, and Applications", Wiley.
8. N. Bhardwaj and S. C. Kundu (2010), *Biotechnol Adv*, **28**, 325.
9. R. Jayakumar, M. Prabaharan, S. V. Nair, and H. Tamura (2010), *Biotechnology Advances*, **28**, 142.
10. S. Y. Chew, Y. Wen, Y. Dzenis, and K. W. Leong (2006), *Current Pharmaceutical Design*, **12**, 4751.
11. J. H. Jang, O. Castano, and H. W. Kim (2009), *Advanced Drug Delivery Reviews*, **61**, 1065.
12. F. Mitschang, H. Schmalz, S. Agarwal, and A. Greiner (2014), *Angewandte Chemie-International Edition*, **53**, 4972.
13. A. Macagnano, E. Zampetti, and E. Kny (2015), "Electrospinning for High Performance Sensors", Springer International Publishing.
14. S. K. Kim (2010), "Chitin, Chitosan, Oligosaccharides and Their Derivatives: Biological Activities and Applications", CRC Press.
15. H. Homayoni, S. A. H. Ravandi, and M. Valizadeh (2009), *Carbohydrate Polymers*, **77**, 656.
16. X. Geng, O. H. Kwon, and J. Jang (2005), *Biomaterials*, **26**, 5427.
17. M. Ignatova, N. Manolova, and I. Rashkov (2013), *Macromolecular Bioscience*, **13**, 860.
18. T. Cheng, R. D. Hund, D. Aibibu, J. Horakova, and C. Cherif (2013), *Autex Research Journal*, **13**, 128.
19. S. De Vrieze, P. Westbroek, T. Van Camp, and L. Van Langenhove (2007), *Journal of Materials Science*, **42**, 8029.
20. K. Ohkawa, D. Cha, H. Kim, A. Nishida, and H. Yamamoto (2004), *Macromolecular Rapid Communications*, **25**, 1600.
21. R. Nirmala, B. W. Il, R. Navamathavan, M. H. El-Newehy, and H. Y. Kim (2011), *Macromolecular Research*, **19**, 345.
22. A. C. Areias, J. A. Gomez-Tejedor, V. Sencadas, J. Alio, J. L. G. Ribelles, and S. Lanceros-Mendez (2012), *Polymer Engineering and Science*, **52**, 1293.
23. Z. M. Mahdieh, V. Mottaghitalab, N. Piri, and A. K. Haghi (2012), *Korean Journal of Chemical Engineering*, **29**, 111.
24. V. Sencadas, D. M. Correia, A. Areias, G. Botelho, A. M. Fonseca, I. C. Neves, J. L. Gomez Ribelles, and S. Lanceros Mendez (2012), *Carbohydrate Polymers*, **87**, 1295.
25. Y. Liu, Y. Liu, N. Liao, F. Cui, M. Park, and H. Y. Kim (2015), *International Journal of Biological Macromolecules*, **79**, 638.
26. K. Sun and Z. H. Li (2011), *Express Polymer Letters*, **5**, 342.
27. Z. M. Huang, Y. Z. Zhang, M. Kotaki, and S. Ramakrishna (2003), *Composites Science and Technology*, **63**, 2223.
28. I. S. Chronakis (2005), *Journal of Materials Processing Technology*, **167**, 283.
29. S. Lin, Q. Cai, J. Ji, G. Sui, Y. Yu, X. Yang, Q. Ma, Y. Wei, and X. Deng (2008), *Composites Science and Technology*, **68**, 3322.
30. G. Wang, Z. Tan, X. Liu, S. Chawda, J. S. Koo, V. Samuilov, and M. Dudley (2006), *Nanotechnology*, **17**, 5829.

31. A. Numnuam, P. Thavarungkul, and P. Kanatharana (2014), *Analytical and Bioanalytical Chemistry*, **406**, 3763.
32. H. Liu, H. Peng, Y. Wu, C. Zhang, Y. Cai, G. Xu, Q. Li, X. Chen, J. Ji, Y. Zhang, and H. W. OuYang (2013), *Biomaterials*, **34**, 4404 .
33. Y. Zhang, J. R. Venugopal, A. El-Turki, S. Ramakrishna, B. Su, and C. T. Lim (2008), *Biomaterials*, **29**, 4314.
34. K. Li, H. Sun, H. Sui, Y. Zhang, H. Liang, X. Wu, and Q. Zhao (2015), *RSC Advances*, **5**, 17541.
35. N. Naseri, C. Algan, V. Jacobs, M. John, K. Oksman, and A. P. Mathew (2014), *Carbohydrate Polymers*, **109**, 7.
36. A. Rothschild, J. Sloan, and R. Tenne (2000), *Journal of the American Chemical Society*, **122**, 5169.
37. E. B. Goldman, A. Zak, R. Tenne, E. Kartvelishvily, S. Levin-Zaidman, Y. Neumann, R. Stiubea-Cohen, A. Palmon, A. H. Hovav, and D. J. Aframian (2015), *Tissue Engineering - Part A*, **21**, 1013.
38. I. Kaplan-Ashiri, S. R. Cohen, K. Gartsman, R. Rosentsveig, G. Seifert, and R. Tenne (2004), *Journal of Materials Research*, **19**, 454.
39. N. Ristolainen, P. Heikkilä, A. Harlin, and J. Seppälä (2006), *Macromolecular Materials and Engineering*, **291**, 114.
40. A. Baklavaridis, I. Zuburtikudis, C. Panayiotou (2017), *J. of Industrial Textiles*, DOI: 10.1177/1528083717725114.

500985 Biodegradable Multilayered Nanofilms for Cell Isolation and Recovery

Abstract:

Selective isolation and purification of circulating tumor cells (CTCs) from whole blood is an important capability for both clinical medicine and biological research. Current techniques to perform this task place the isolated cells under excessive stresses that reduce cell viability, and potentially induce phenotype change, therefore losing valuable information about the isolated cells. The goal of our work is to effectively isolate as well as non-inversely recover cancer cells using a microfluidic device coated with a biodegradable multilayered nanofilm. To this end, we have applied layer-by-layer (LbL) assembly to create a library of ultrathin coatings using a broad range of materials through complementary interactions. We systematically studied the effect of various flow conditions and channel geometries on the thickness and surface roughness of the resulting films. We also investigated the biocompatibility and degradation behaviors of a series of enzymatically-degradable films made from naturally derived polymers. By developing an LbL nano-film coating with an affinity-based cell-capture surface that is capable of selectively isolating cancer cells from whole blood, and that can be rapidly degraded on command, we are able to gently isolate cancer cells and recover them without compromising cell viability or proliferative potential. This film system has been applied to two capture and release platforms: 1) microfluidic HB chip and 2) hollow glass microspheres. Detailed characterization on the film system was conducted via fluorescent microscopy, AFM, profilometer, TGA, QCM-D, immunofluorescent staining, etc. Our approach has the capability to overcome practical hurdles and provide viable cancer cells for downstream analyses.

17AIChE
Annual Meeting, Minneapolis, MN

Ag/Cu Bimetallic Nanoparticle- and Ionic-Graphene Composites with Enhanced Antibacterial Performance

Anna Perdikaki[1], Angeliki Galeou[2], George Pilatos[1], Anastasia Prombona[2], Georgios N. Karanikolos[3*]

Institutes of [1]Nanoscience and Nanotechnology (INN), and [2]Biosciences and Applications, Demokritos National Research Center, Athens, Greece

[3]Department of Chemical Engineering, Petroleum Institute, Khalifa University of Science & Technology, Abu Dhabi, UAE

Continuous Need for More and More Efficient Antimicrobial Agents

- ❑ Increased proliferation of antimicrobial-resistant and new strains of bacterial pathogens.
- ❑ Severe impact on current health, environmental, and technological developments.
- ❑ Demand for design of novel, highly efficient antibacterial agents.

Antibacterial action of metallic species →

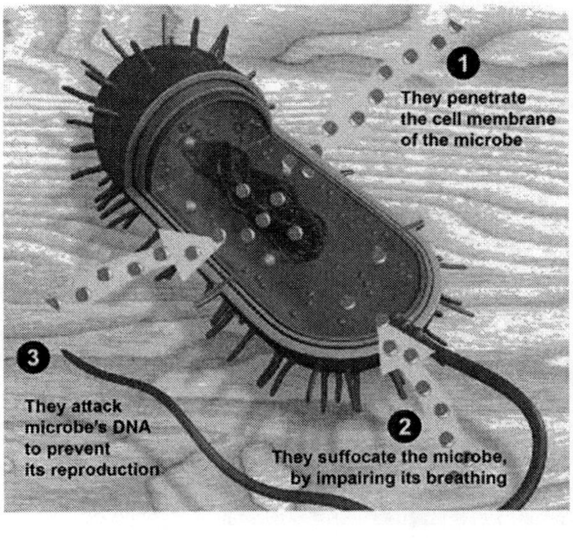

Focus of this work:

- Use of graphene as support for metal NPs and ions.
- Employ bimetallic species populations.

Growth of Few-Layer Graphene as Support

✓ Graphene–based support was grown by CVD using ferrocene – *a common precursor for CNT growth* - as the only catalyst/carbon precursor

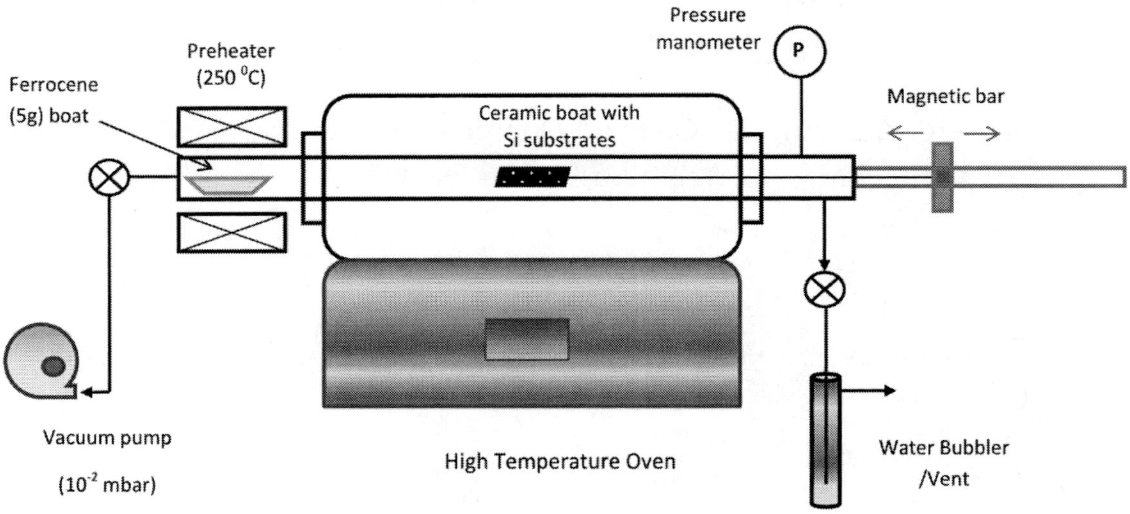

Graphene Support Properties

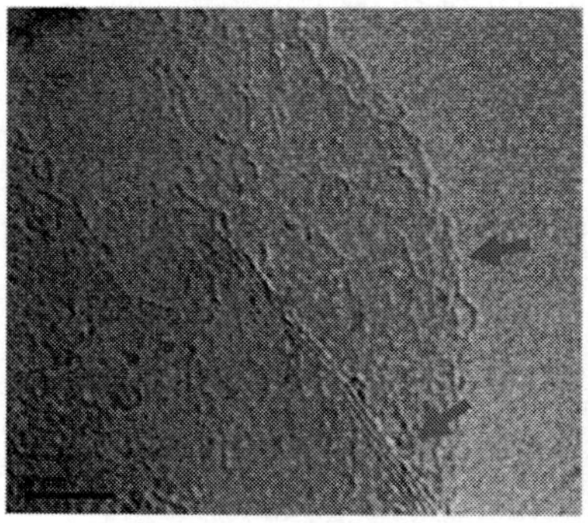

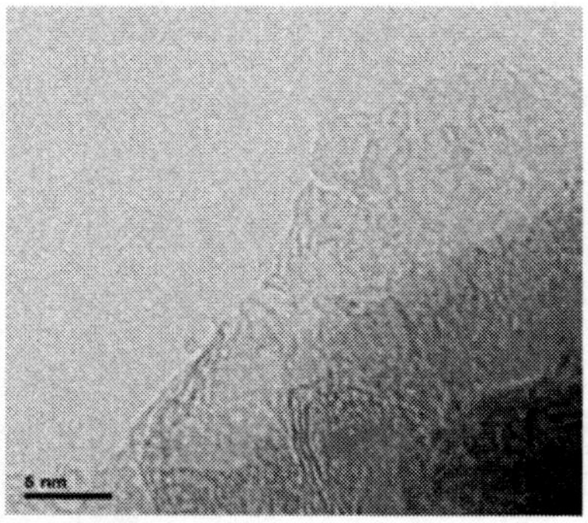

Support based on 2-3 layer graphene

Raman:

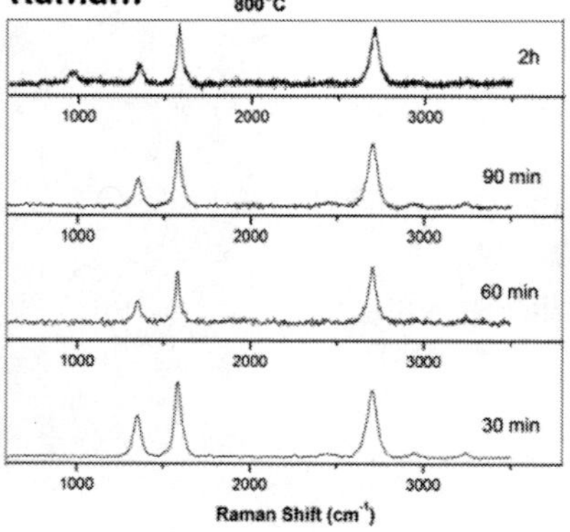

Full Width at Half Maximum (FWHM) of 2D Raman band at ~49 cm^{-1}

Porous Support - Preferred Morphology

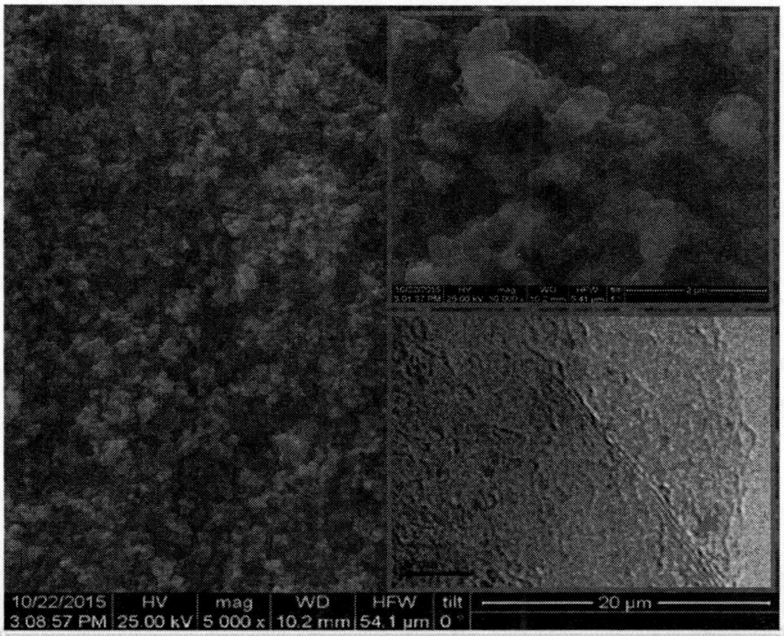

Irregularly shaped particles forming a tortuous network

✓ **Enhance dispersion of metallic nanoparticles and interaction with the graphene**

✓ **High surface area → Facilitate exposure of the bacteria**

J. App. Phys. 2016

Oxidation/Functionalization of Graphene Support

1. **Post-treatment of the graphene-based support using an acidic HNO_3/H_2SO_4 solution to introduce oxygen functional groups on the surface**
2. **Growth-on-surface and stabilization of the metallic NPs**

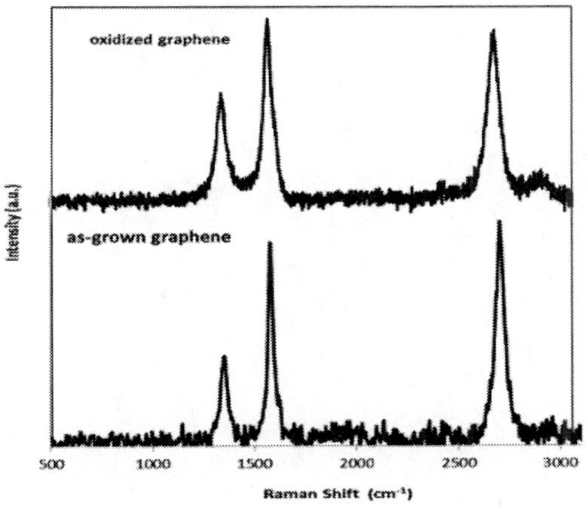

Oxidation did not affect the graphene structural characteristics significantly

- I_{2D}/I_G: **1.1** (as-grown) → **0.9** (oxidized)
- I_D/I_G: **0.4** (as-grown) → **0.6** (oxidized)

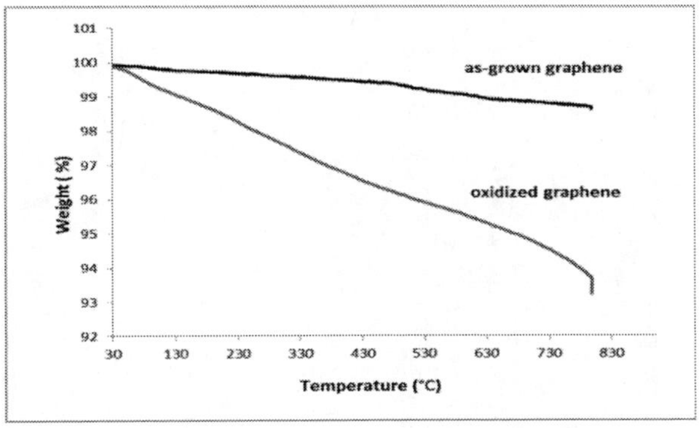

TGA:

- ~100 °C: release of water
- 200-350 °C: decomposition of the oxygenated functional groups
- Higher temperatures: pyrolysis of residual oxygen-containing groups and carbon skeleton of graphene oxide

Graphene and Oxidized Graphene Effect on Bacterial Culture *(E. Coli)*

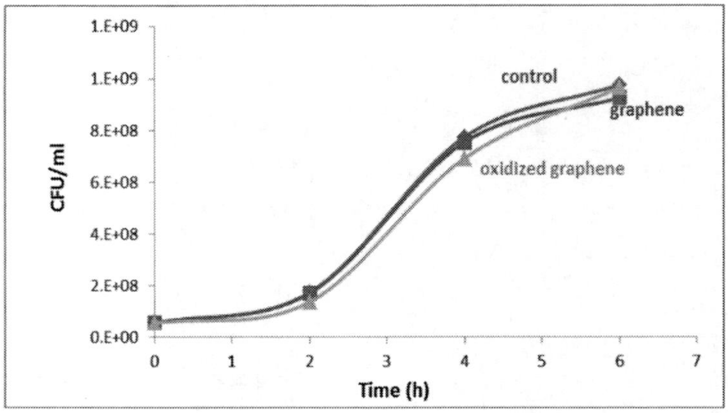

- No significant antibacterial activity of the graphene-based support against the *E.coli* culture.
- This allows us to attribute any antibacterial activity exclusively on the supported metal species.

Photographs of the agar plates showing the bacterial colonies after a 6h incubation period

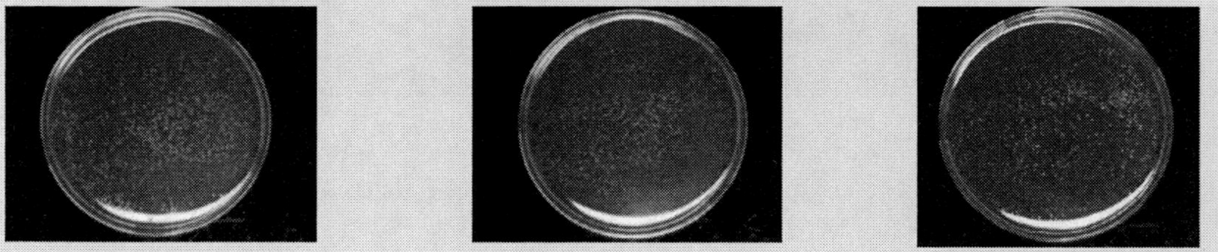

control **graphene** **oxidized graphene**

In situ metallic NP growth

Functionalized graphene with attached Ag^+ ions

AgNPs - graphene hybrid by in - situ growth

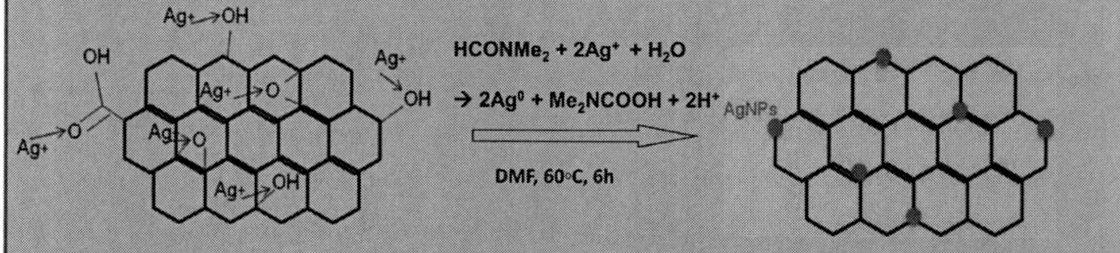

✓ Oxygen functional groups (epoxy, hydroxyl, and carboxyl groups) stabilized Ag^+ ions by electrostatic interactions

Uniform metal distribution and *in situ* generation of metallic nanoparticles

Surface-bound ions → Surface-bound metal NPs

Metal NP-Graphene Hybrid

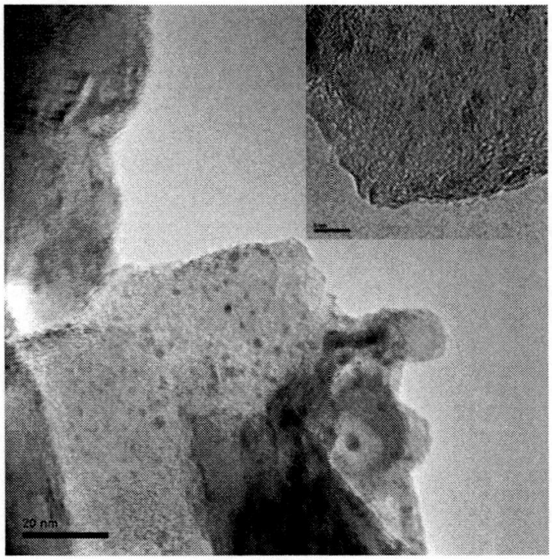

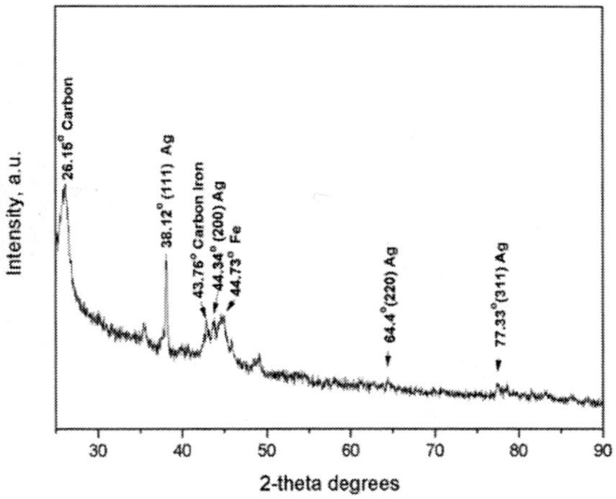

- Particles of less than 3 nm evident
- Diffraction peaks at 2-theta of 38.12°, 44.34°, 64.40° and 77.33° from (111), (200), (220) and (311) planes of crystalline Ag, respectively
- The graphene peak is at 26.15°

Antibacterial Activity of AgNP-Graphene Hybrid

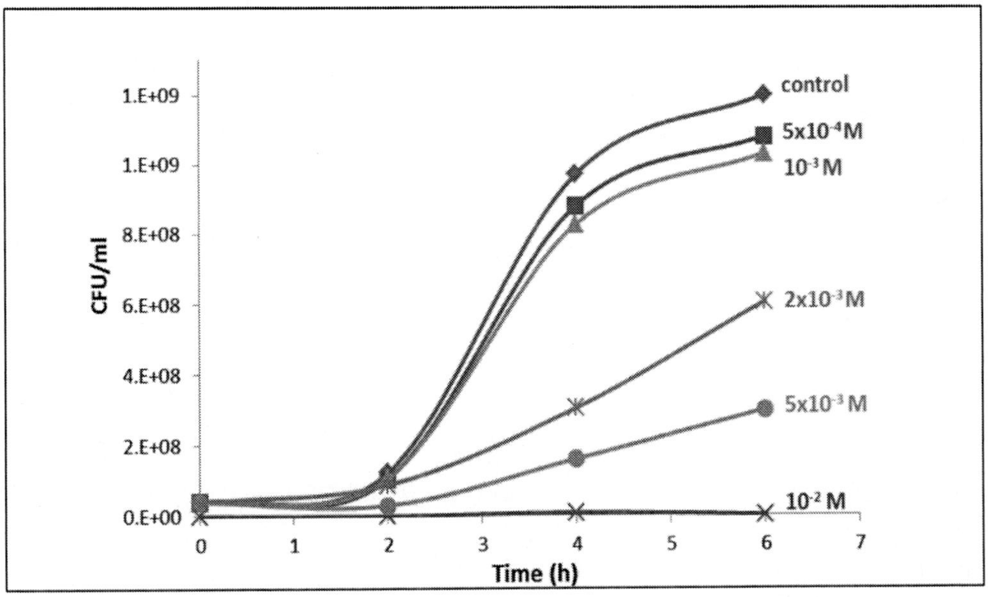

- ➢ **Bacterial growth is suppressed as metal concentration increases**
- ➢ **Metal concentration of 10^{-2} exhibits bactericidal effect**

AgNP-Graphene Hybrid vs. Colloidal Ag NPs

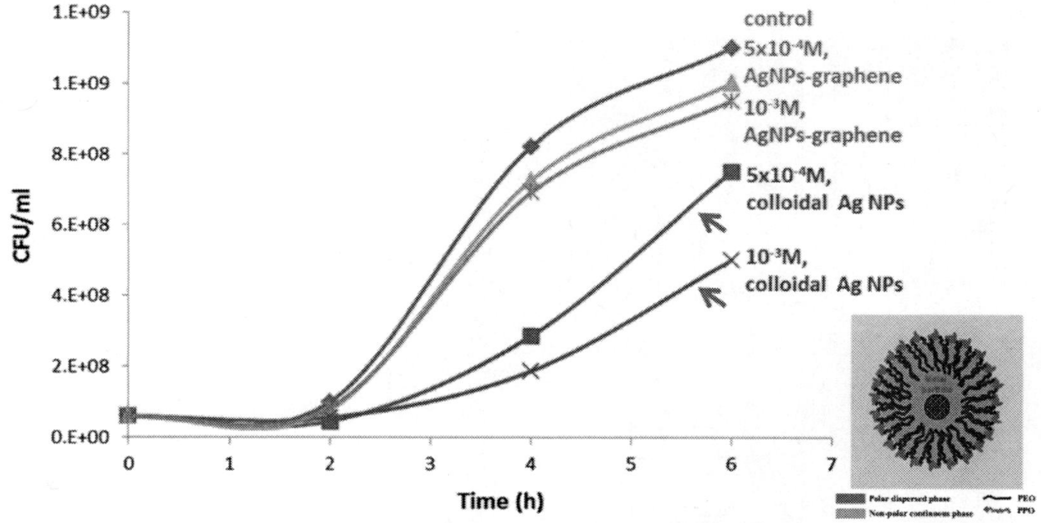

- For comparison, colloidal AgNPs were grown in solution and stabilized by a block copolymer *(G.N. Karanikolos et. al., Langmuir 2013)*.
- Colloidal AgNPs antibacterial activity was higher than that of the AgNP-graphene hybrid, with same metal concentrations.

Colloidal AgNPs :

Well-dispersed → Isolated due to block copolymer capping → limited aggregation.

CuNP-Graphene Hybrid

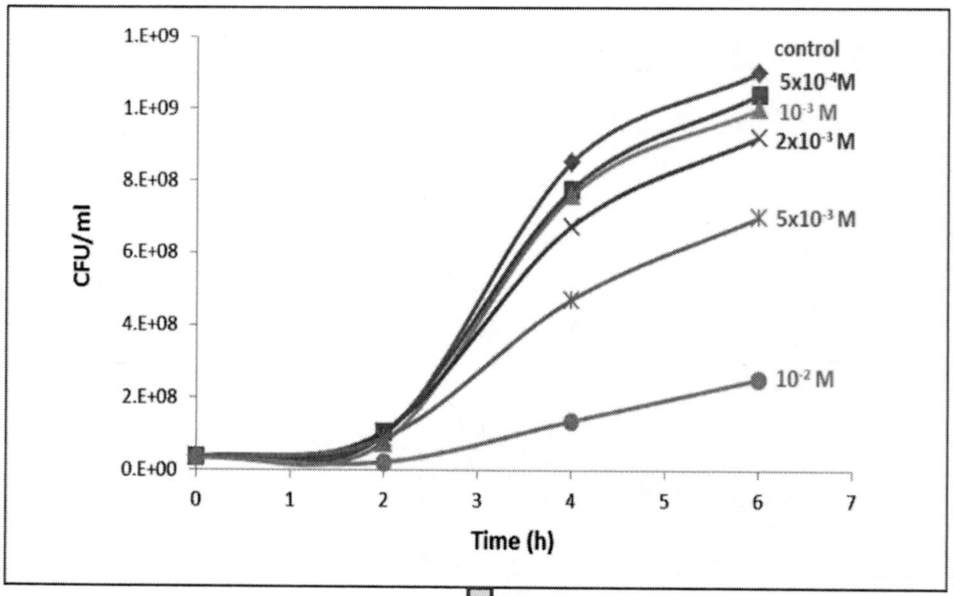

- **Bacterial growth is suppressed as metal concentration increases.**
- **CuNP-Graphene hybrids less active compared to the AgNP ones.**

Antibacterial effect of CuNP-Graphene vs. AgNP-Graphene

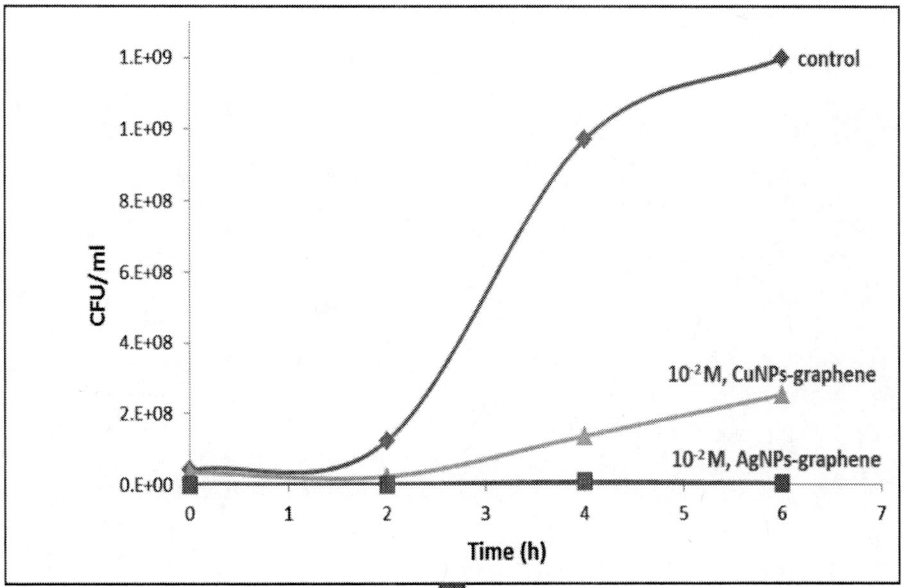

→ The AgNP-Graphene hybrid is more active compared to the CuNP one

Bimetallic Antibacterial effect of Ag/CuNP-Graphene Hybrid

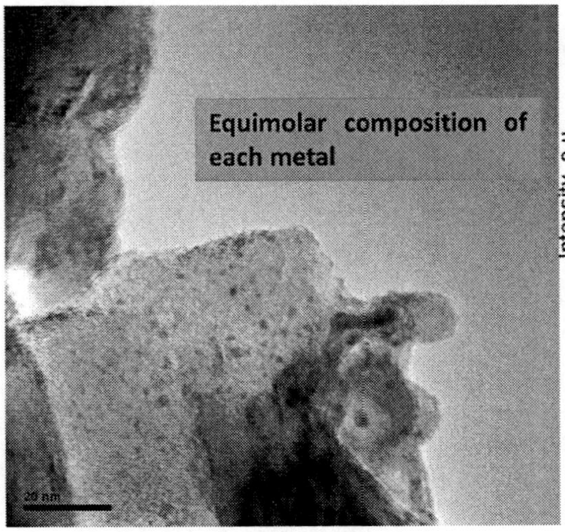

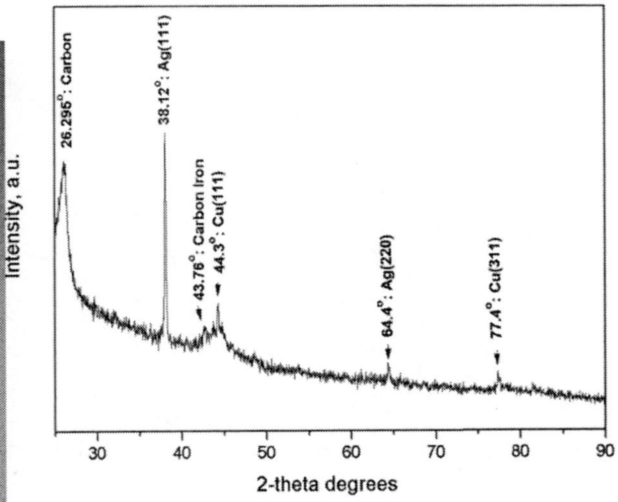

TEM:

- Average NP size ≈3 nm

XRD:

- 38.12° and 64.4° → (111) and (220) crystallographic planes of Ag, respectively
- 44.3°, 77.4° → (111) and (311) crystallographic planes of Cu, respectively

Surface Analysis of Ag/CuNP-Graphene Hybrid

X-ray elemental distribution of Ag (green) and Cu (blue) in Ag/CuNP–graphene hybrid (total metal concentration of 5 × 10^{-3} M)

✓ Coexistence of both Ag and Cu in the bimetallic Ag/CuNP–graphene material

✓ Ag exhibits a more uniform and homogeneous distribution

Antibacterial effect of bimetallic Ag/CuNP-Graphene Hybrid vs. monometallic ones on *E. Coli* cells

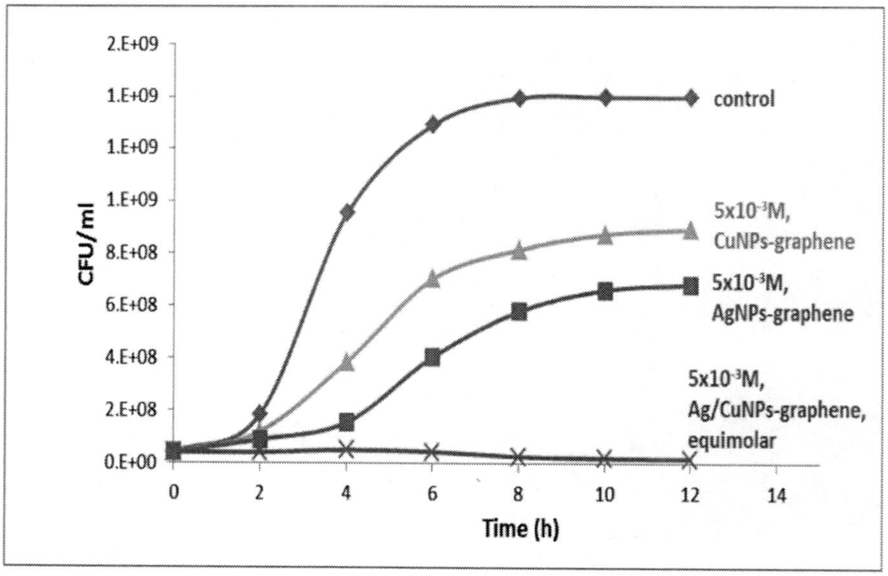

- **Bimetallic Ag/CuNP-graphene hybrid exhibits significantly higher antibacterial activity compared to monometallic ones of same metal concentration**
- **Stable action even after prolonged incubation period (12 h)**

ACS Appl. Mater. Interf. 2016

Antibacterial effect of bimetallic Ag/CuNP-Graphene Hybrid vs. monometallic ones on *Synechococcus* cells

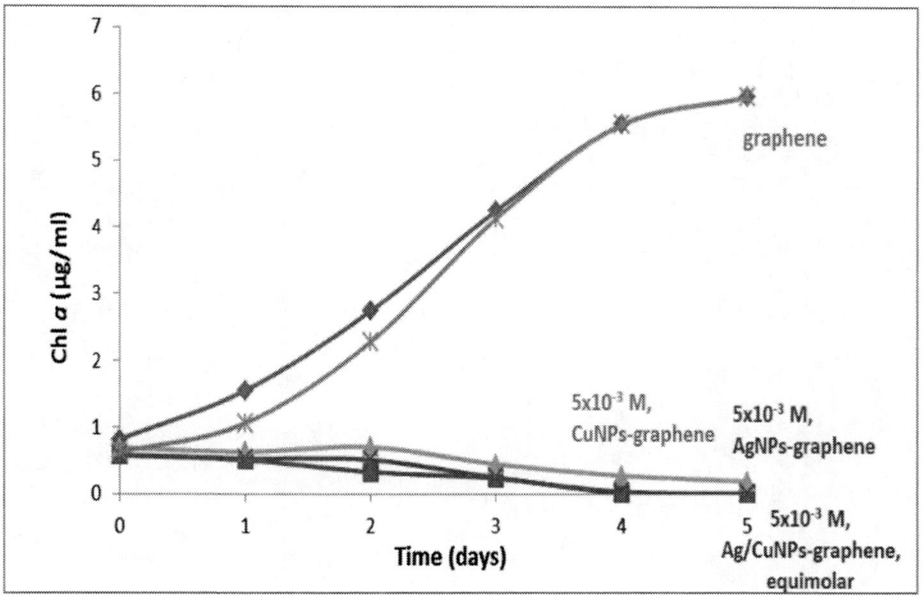

Bimetallic Ag/CuNP-Graphene Hybrid vs. Colloidal NPs

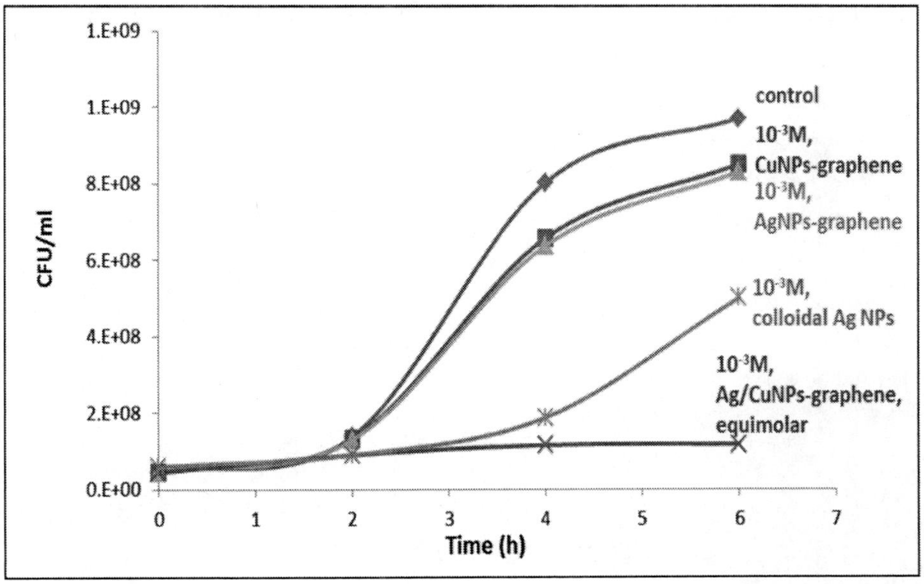

- ✓ The bimetallic Ag/CuNP-graphene hybrids demonstrate higher antibacterial performance compared to colloidal metal NPs

Bimetallic Ag/CuNP-Graphene Hybrid vs. Monometallic ones and Colloidal NPs

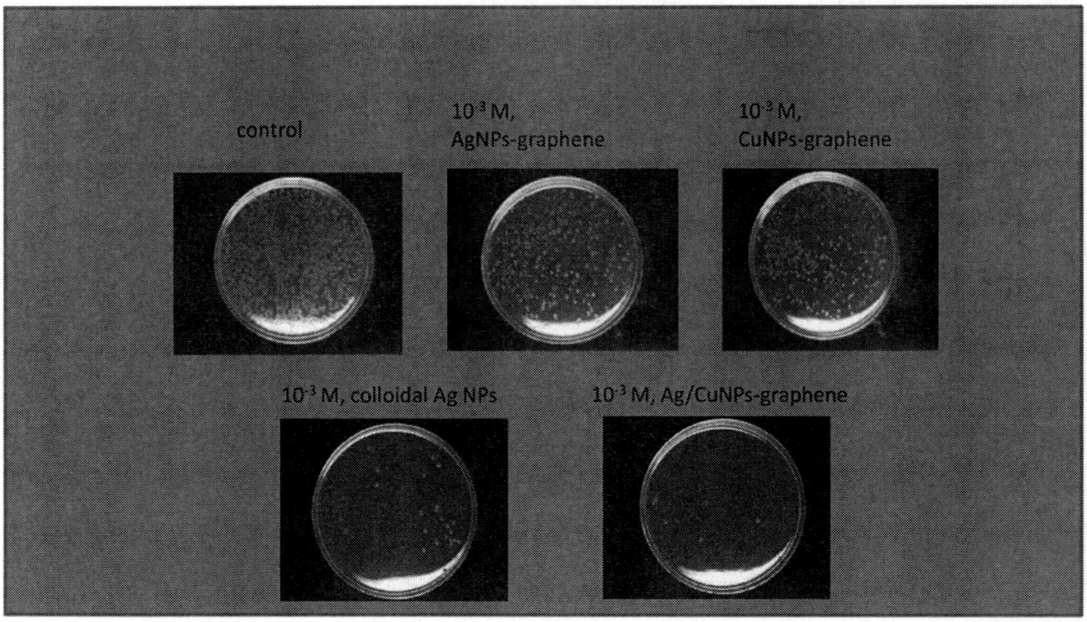

Photos of the *E. Coli* bacterial colonies in agar plates after a 6 h incubation period

Purely Ionic Structures

Ion-based, metal/graphene hybrid analogues comprising of surface-bound Ag, Cu mono-ionic and Ag/Cu bi-ionic species on functionalized graphene, without involvement of NPs:

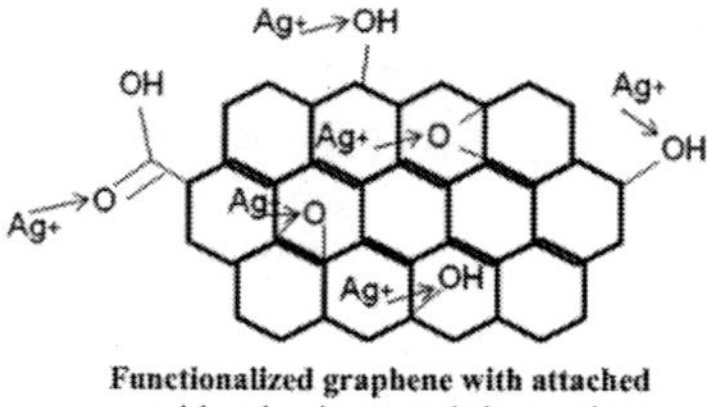

Functionalized graphene with attached metal ions by electrostatic interactions

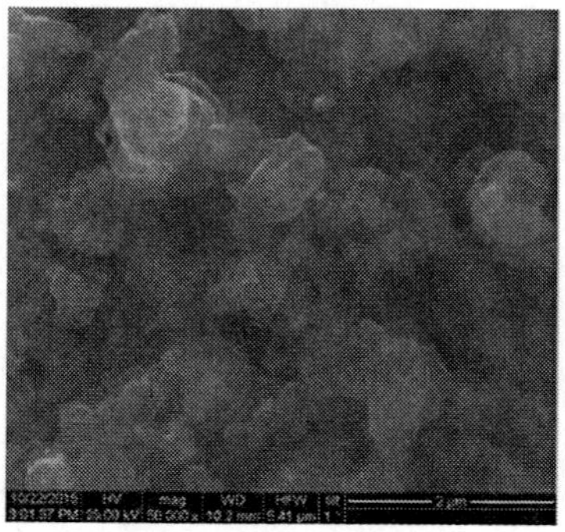

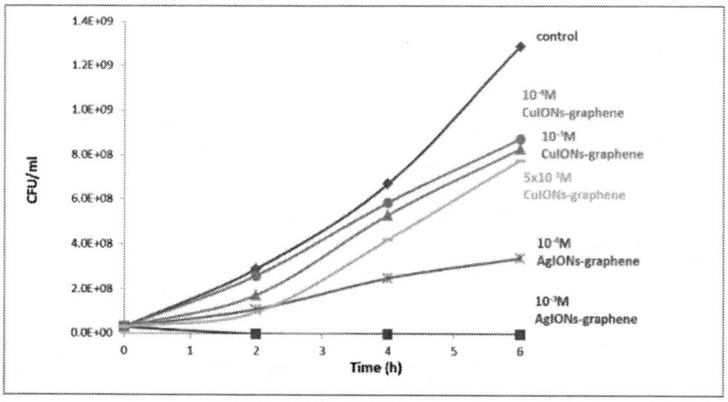

- The ionic structures exhibit high antibacterial activity as well.
- The AgIONs-graphene material is more efficient than the CuIONs-graphene one - similarly to the NP-based structures.

Purely Ionic Structures: Bimetallic vs. Monometallic

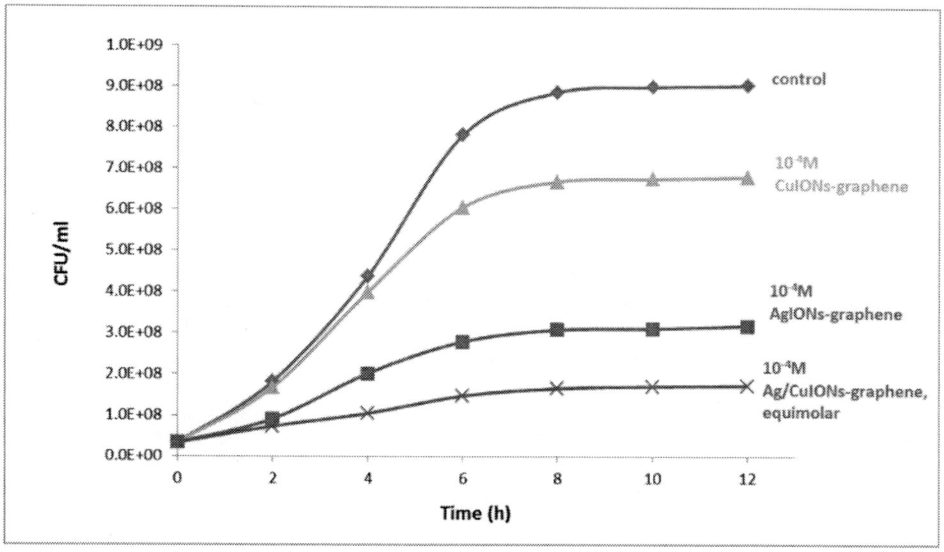

- ✓ The bimetallic Ag/CuIONs-graphene hybrids demonstrate higher antibacterial performance compared to the monometallic ones – analogous behavior as with the NP-based materials
- ✓ Activity unchanged even after extended incubation period (12h)

Purely Ionic Structures: Ion- vs NP-based systems

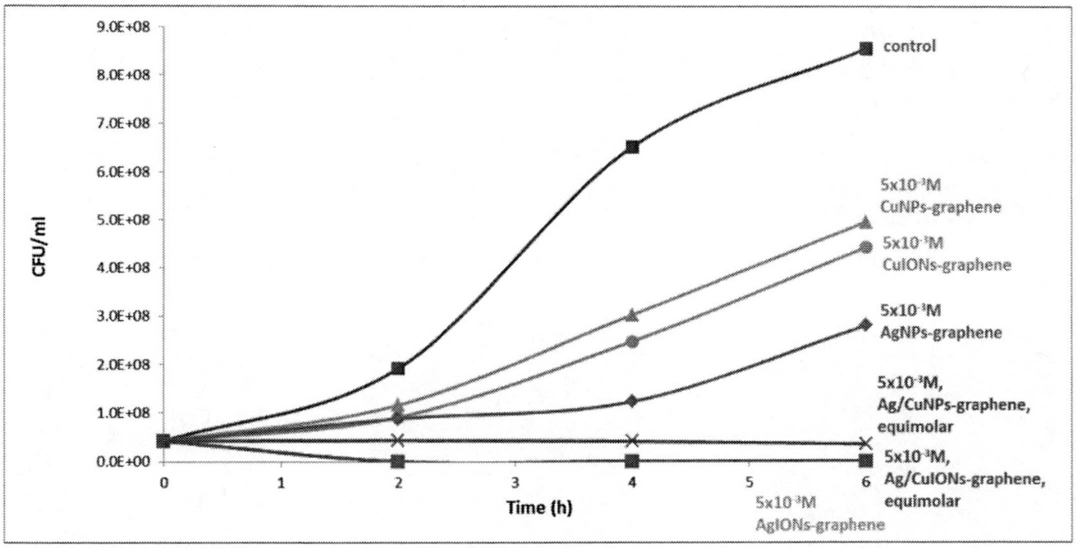

- ✓ The ionic systems are more active than the NP-based ones of same metal type and concentration

Stability of the ion-based systems in solution

Materials underwent washing cycles before the bacterial tests:

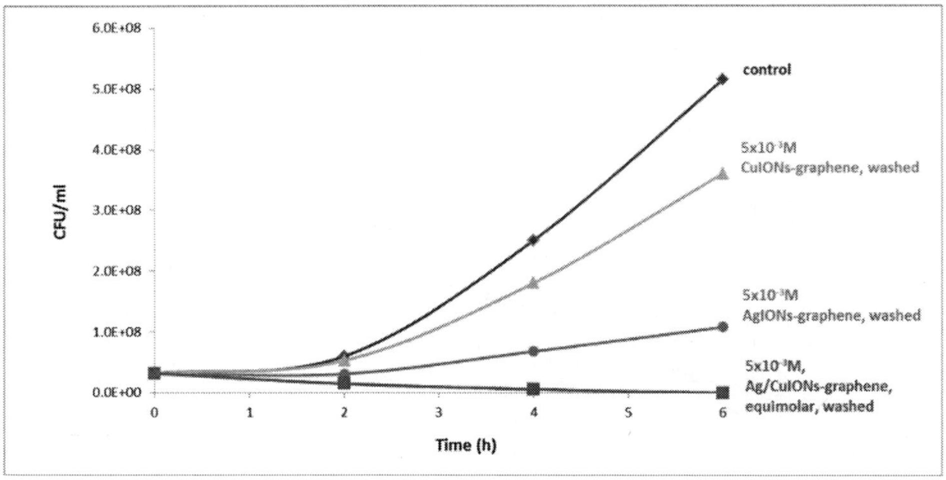

✓ The developed ionic systems are stable, as their performance remained almost unchanged even after washing the material repeatedly before executing the antibacterial testing, indicating that the ions are well attached on the surface functional groups of graphene and do not migrate into the liquid phase, i.e. acting directly from the surface of the graphene particles.

Agar plates after 6 hours of incubation

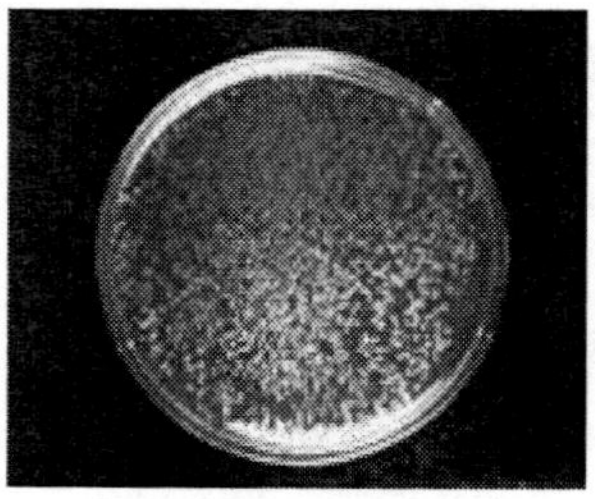
control

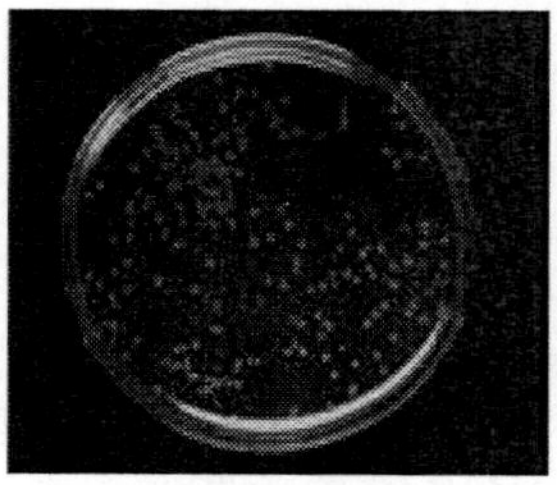
$5x10^{-3}$ M, CuNPs-graphene

$5x10^{-3}$ M, AgNPs-graphene

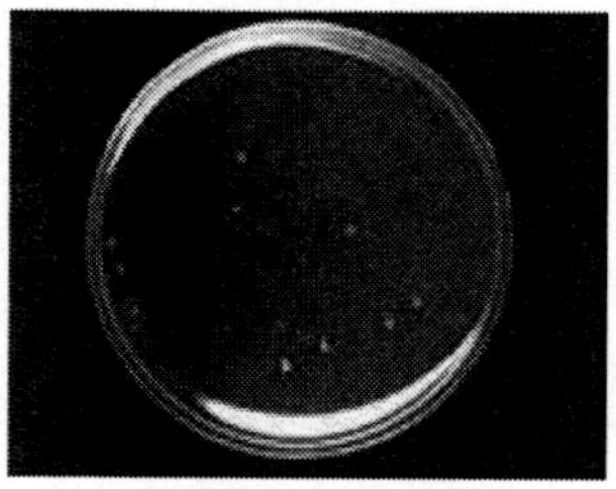
$5x10^{-3}$ M,
Ag/CuNPs-graphene,
equimolar

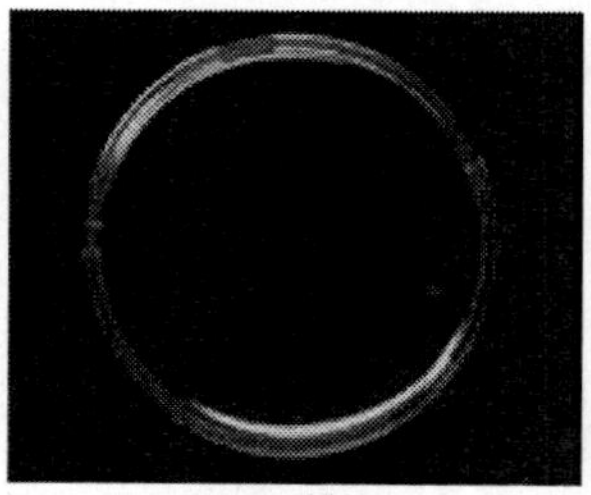
$5x10^{-3}$ M,
Ag/CuIONs-graphene,
equimolar

$5x10^{-3}$ M,
Ag/CuIONs-graphene,
equimolar, washed

Cyclability/reusability of the ionic systems

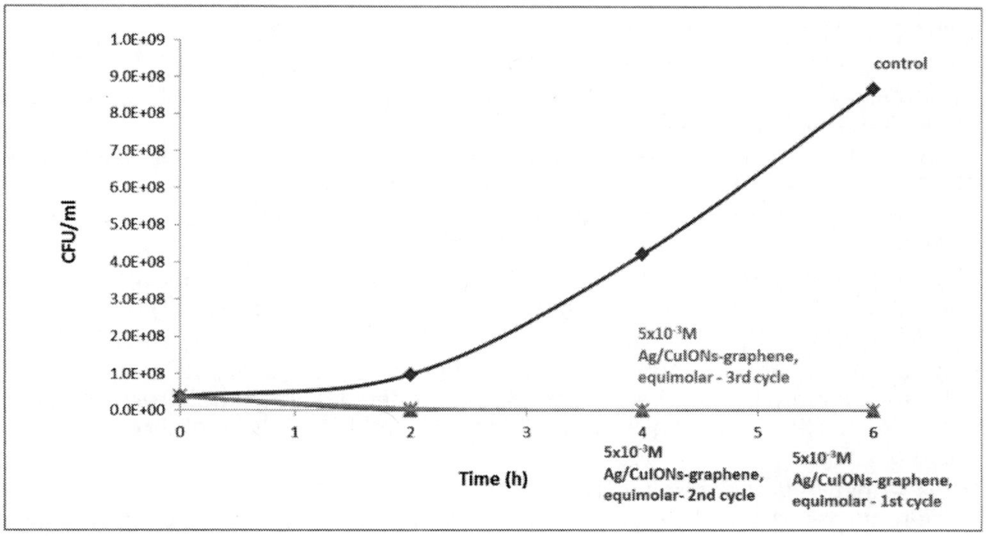

- ✓ The hybrids are reusable as their high activity was maintained after exposure to repeated cycles of bacterial incubation using fresh bacteria each time.

Agar plates after 6 hours of incubation

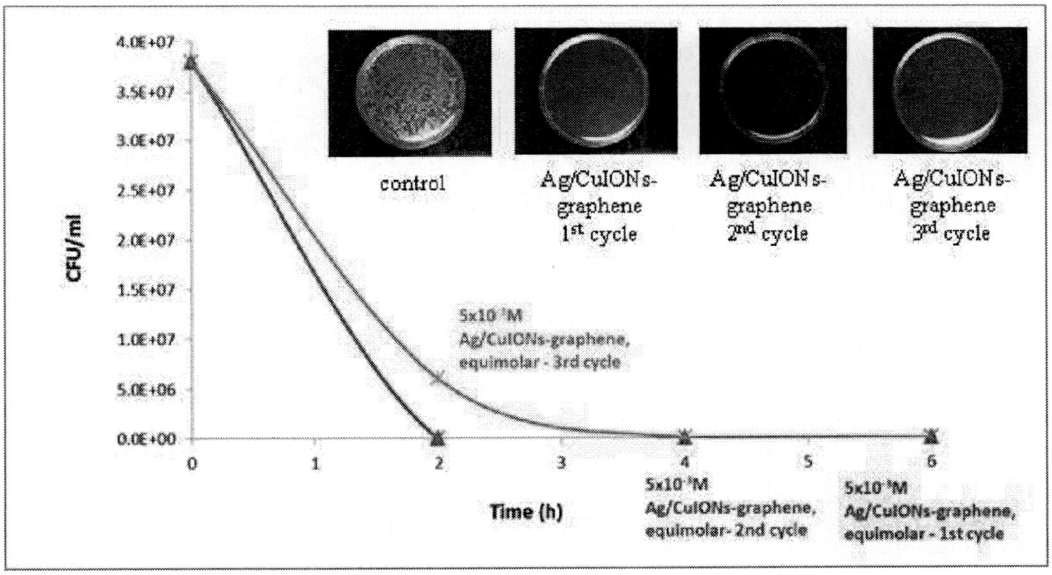

Conclusions

- Ag, Cu monometallic and Ag/Cu bimetallic NP- and ionic-based graphene antibacterial hybrids were developed.
- Porous graphene-based support was grown by CVD using ferrocene as catalyst/carbon precursor, and further functionalized by sonochemical oxidation as to introduce oxygen functional groups on the surface.
- Tortuous/porous morphology of graphene support → Trapping of the bacteria cells in its framework → Direct contact of the bacteria cells with the deposited metallic species → Locally high concentration of ions → Significant inhibition of bacterial growth.
- The Ag- and Cu- based monometallic graphene hybrids significantly suppress bacterial growth, yet the Ag-graphene ones exhibit higher activity compared to their Cu-based counterparts.
- The bimetallic Ag/Cu-graphene hybrids exhibit superior performance compared to all other materials tested, i.e. both the monometallic graphene structures as well as colloidal NPs, achieving complete bacterial inhibition in all metal concentrations tested.
- This enhanced performance is attributed to the synergistic action of the two different metals that coexist on the surface as well as the enhancing role of the graphene support.
- The ion-based materials exhibit higher activity than the NP-based analogues.
- The ionic systems exhibit stability in solution, and recyclability/sustained performance as their activity was preserved after repeated cycles of bacterial exposure.

17AIChE

THANK YOU

Fabrication of homologous TiO_2/NH_2-MIL-125(Ti) heterojunction for efficient visible-light-induced aerobic oxidation of benzyl alcohol and degradation of tetracycline

Xiyi Li, Yunhong Pi, Hao Yu, Zhong Li and Jing Xiao*

School of Chemistry and Chemical Engineering
South China University of Technology

30 October, 2017, AIChE, Minneapolis, America

Outline

1. Background
2. Research objective
3. Experimental and discussion
4. Conclusions

AIChE
The Global Home of Chemical Engineers

PART ONE 1 Background

AIChE
The Global Home of Chemical Engineers

Background | Objective | Experiment | Conclusion

4

MOFs

- As a new class of versatile porous solid materials, MOFs show promises to be tuned as effective photocatalysts via post-synthetic modification (PSM).

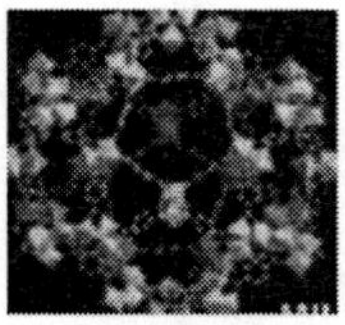

High specific surface area & Pore volume

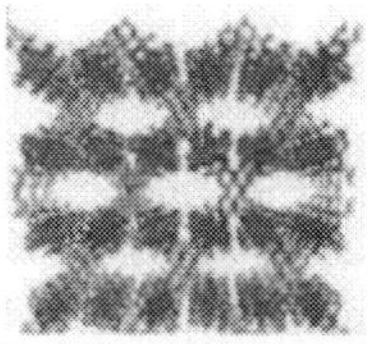

Tailorable pore structure & surface properties

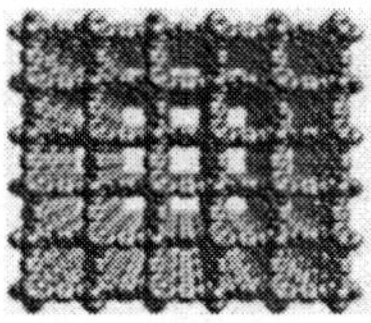

Uniform pore size distribution

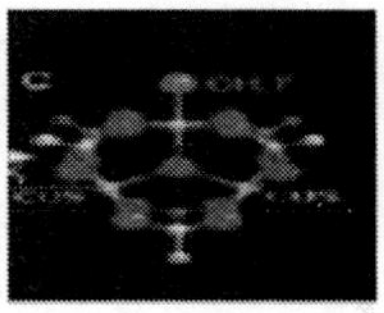

Diverse ways of coordination

5

PART TWO

2

Objective

AIChE The Global Home of Chemical Engineers

Background | Objective | Experiment | Conclusion

1 The integration of MOFs photocatalyst with TiO_2 materials

2 Clarification of the structure and property

3 Investigation of the photocatalytic activity

4 Clarification of the mechanistic pathways

7

AIChE The Global Home of Chemical Engineers

PART THREE 3 Experiment

AIChE

The Global Home of Chemical Engineers

Background | Objective | Experiment | Conclusion

◆ TiO_2/NH_2-MIL-125(Ti)(LP-3) composite was synthesized through a facile method.

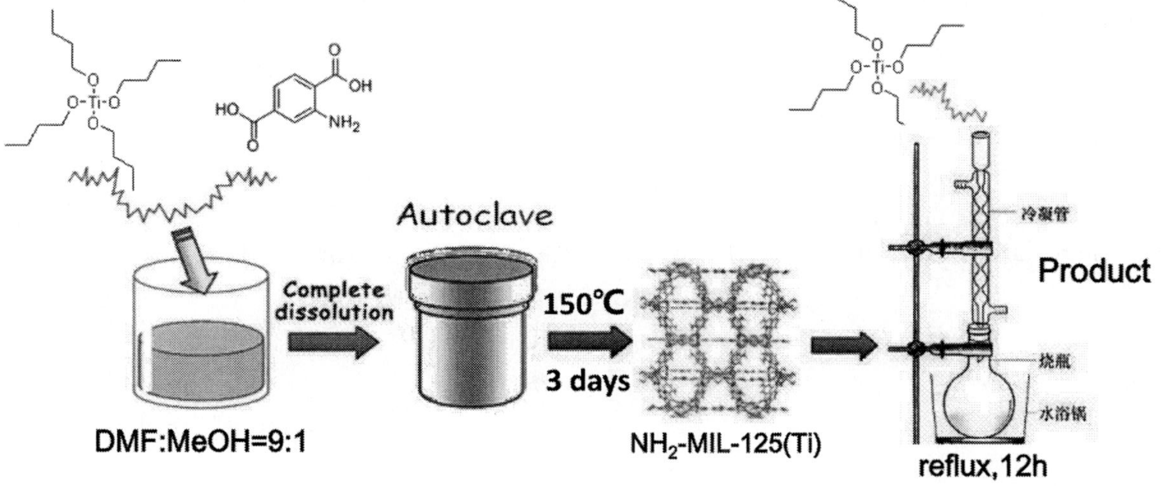

Synthesis process

Characterizations of TiO_2/NH_2-MIL-125(Ti)

◆ **XRD**

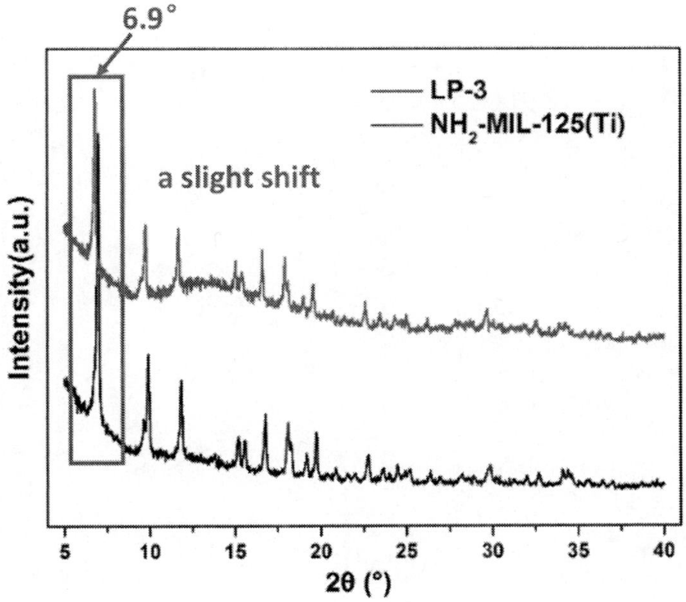

The framework of NH_2-MIL-125(Ti) keeps static after the incorporation of TiO_2.

Characterizations of TiO_2/NH_2-MIL-125(Ti)

◆ N_2 adsorption-desorption

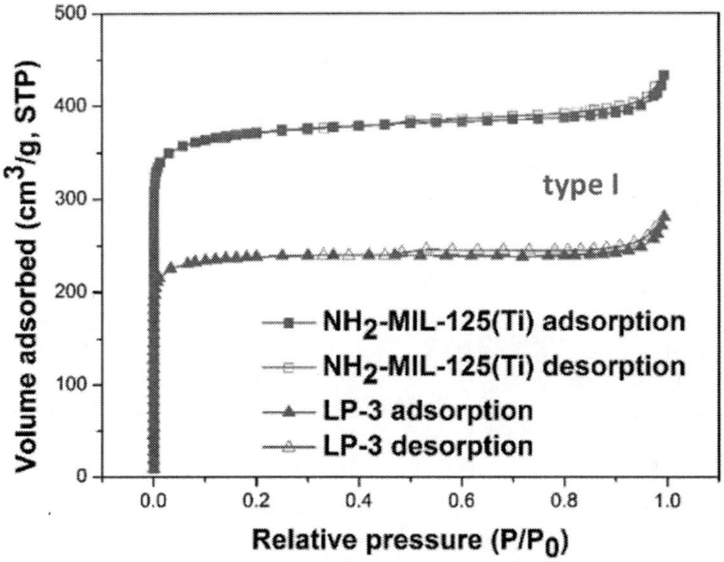

Similar behavior of the LP-3 revealed that the pore structure stayed intact.

11

Characterizations of TiO_2/NH_2-MIL-125(Ti)

◆ SEM & TEM

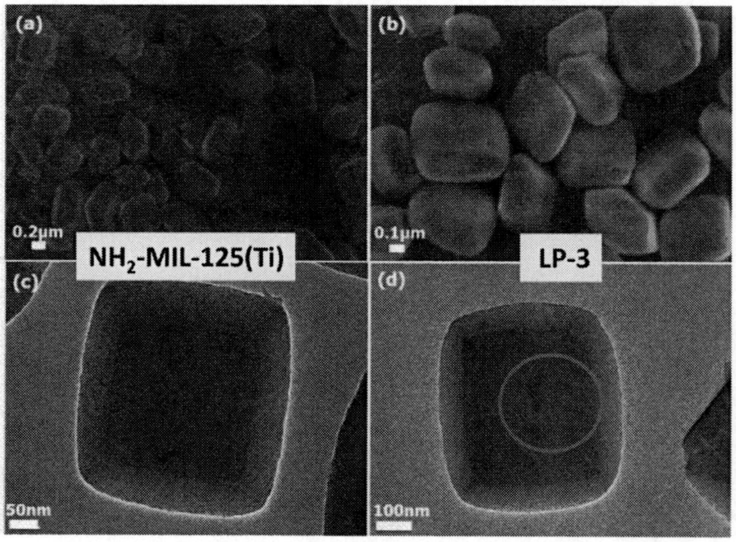

The unchanged morphology of LP-3 indicates that hybridization does not alter the basic crystallinity of NH_2-MIL-125(Ti).

12

Characterizations of TiO_2/NH_2-MIL-125(Ti)

◆ **XPS**

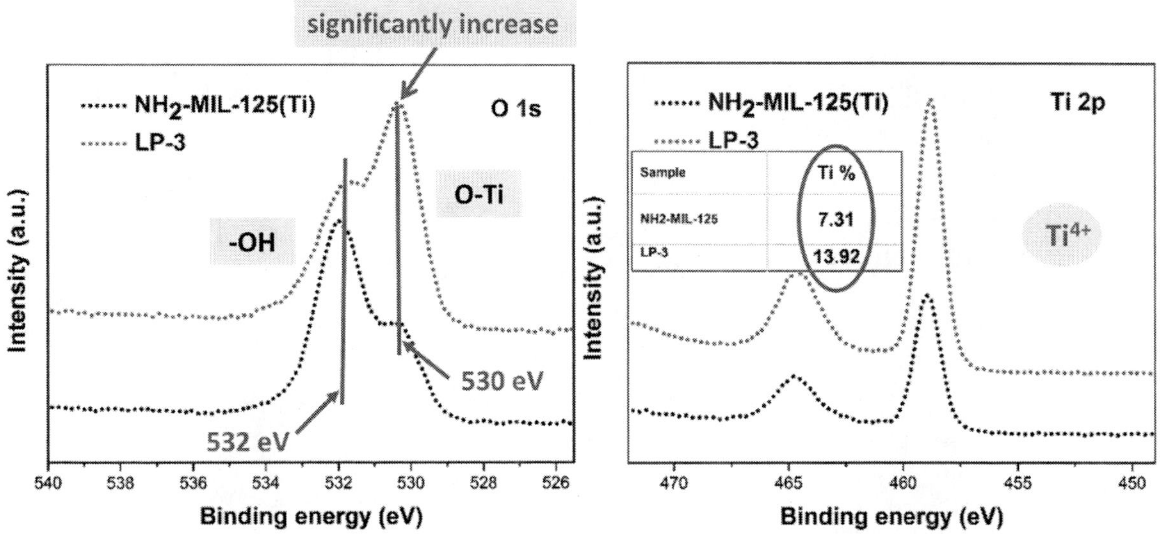

the successful incorporation of TiO_2 **in LP-3**

13

Performance of LP-3

◆ **Degradation of tetracycline**

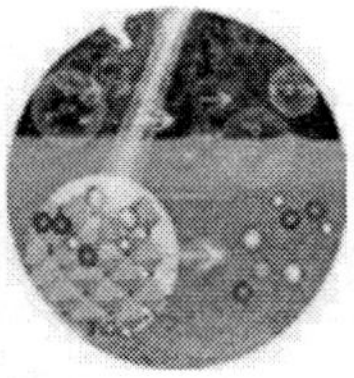

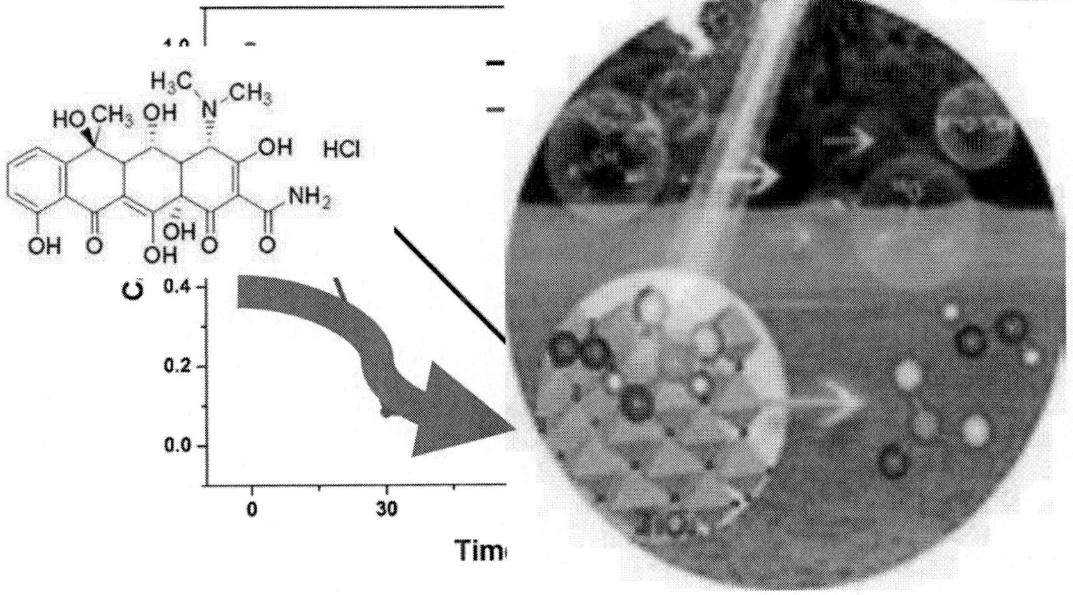

15

Performance of LP-3

- Photocatalytic oxidation of benzyl alcohol

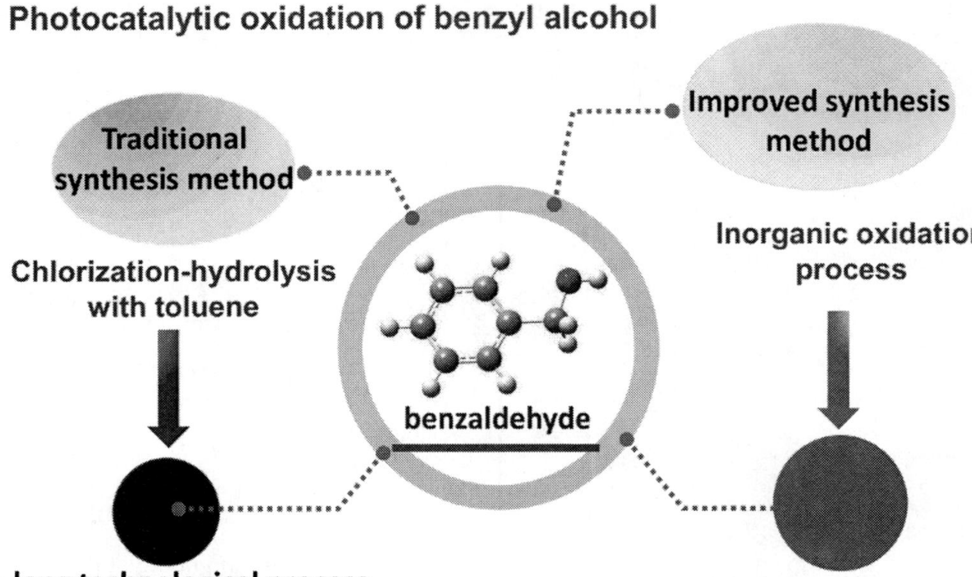

- long technological process
- Difficulty of product separating
- Discharge of caustic gas Cl_2

- Reagent waste
- Environmental pollution

16

Performance of LP-3

- Photocatalytic oxidation of benzyl alcohol

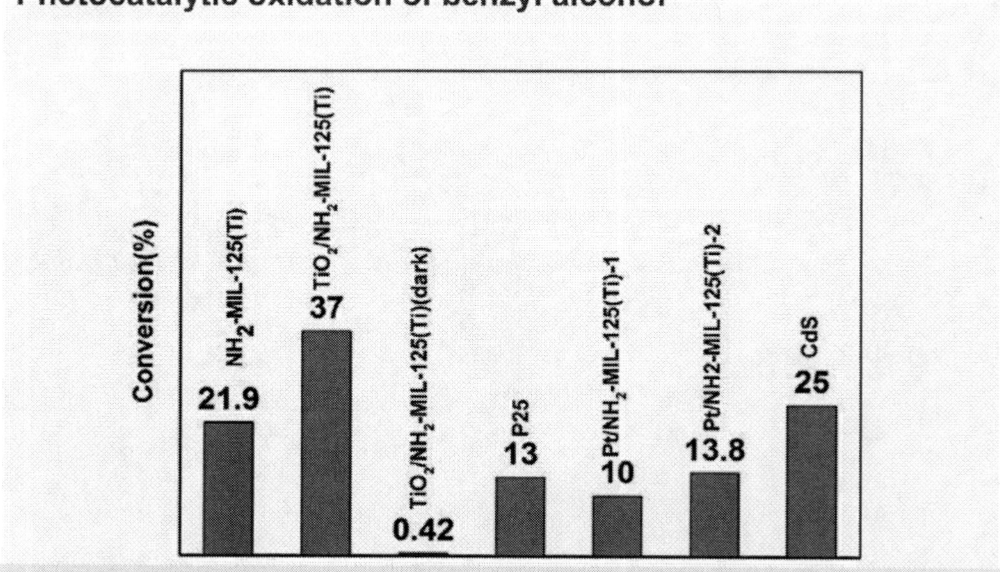

The Conversion of photocatalytic oxidation of benzyl alcohol.
(Reaction condition: 50 mg photocatalyst, 6 ml solvent, 3 mmol benzyl alcohol, 1 atm O_2, and 300 W Xenon lamp >420 nm)

17

Reusability of LP-3

high reusability and stability

◆ Regeneration cycles & XRD & SEM

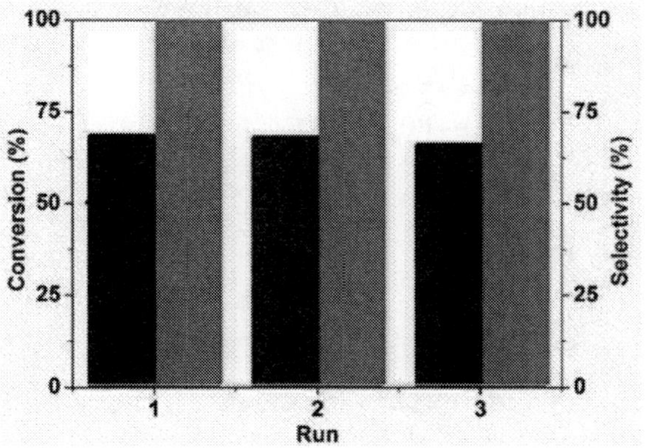

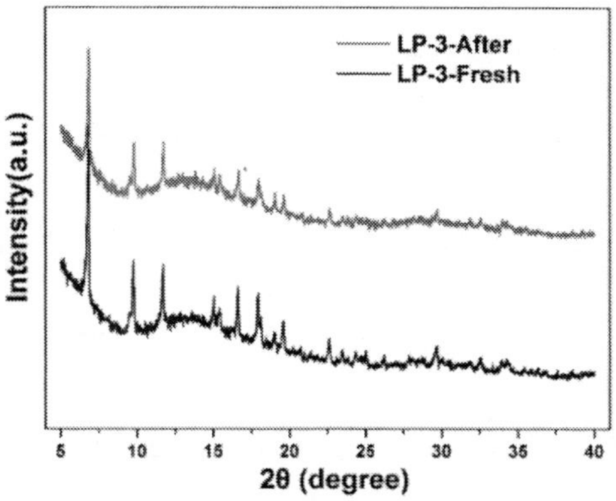

before

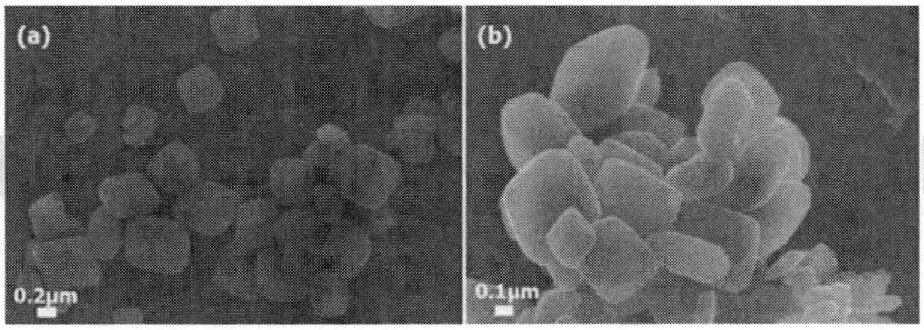

after

18

Characterization of mechanism on LP-3

larger amount of generated $O_2^{\cdot -}$

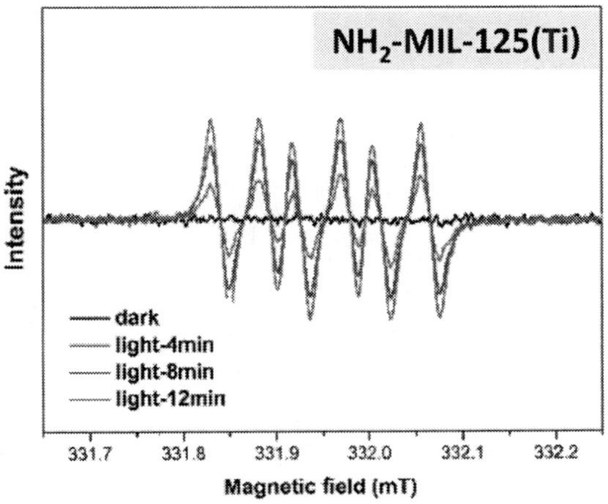

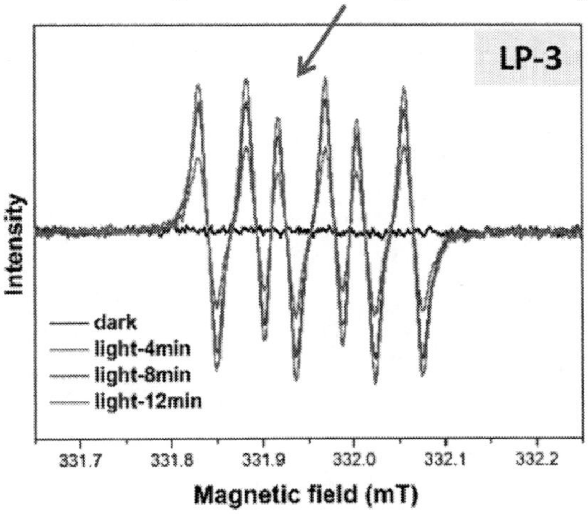

ESR spectra of $O_2^{\cdot -}$

19

Characterization of mechanism on LP-3

Photocurrent response

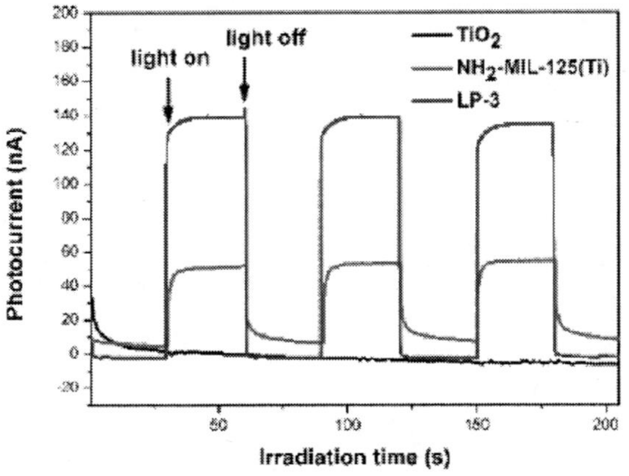

EIS spectra

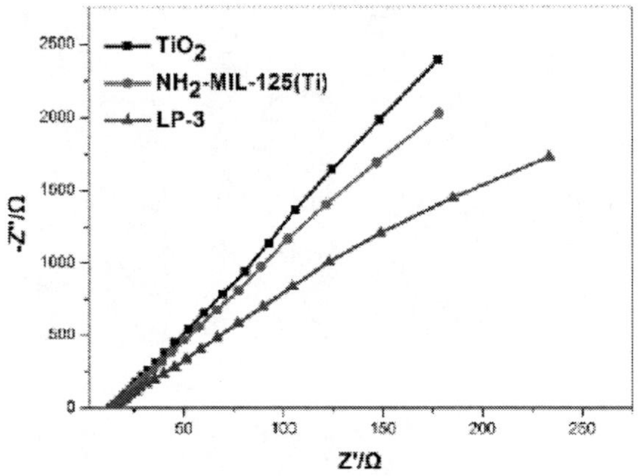

Charge transfer was facilitated by incorporation of TiO_2 in the LP-3 composite, resulting in the decreased recombination of photo-induced e^-/h^+, leading to higher conversion of benzyl alcohol.

20

Characterization of mechanism on LP-3

ESR spectra with g values

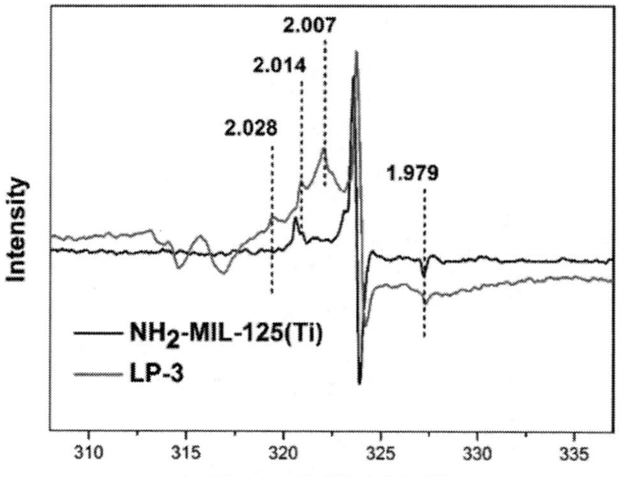

After the efficient transfer of photo-induced e-/h^+, different from Pt/NH_2-MIL-125(Ti), LP-3 can preserve Ti^{3+} well for the activation of O_2 molecule to $O_2^{\cdot -}$ immediate for benzyl alcohol oxidation.

21

Possible mechanism on LP-3

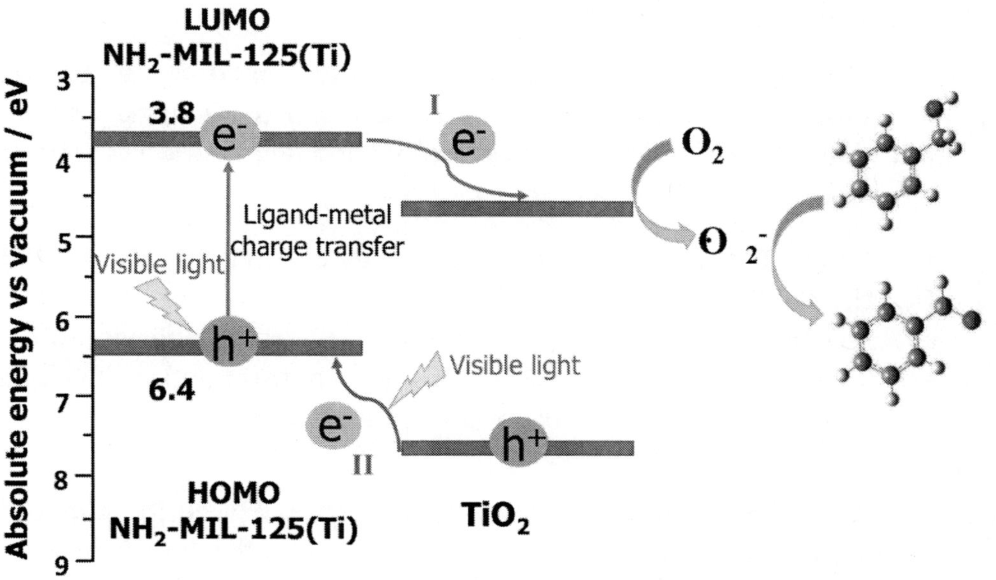

Schematic illustration of the photocatalytic oxidation over LP-3

22

AIChE
The Global Home of Chemical Engineers

PART FOUR

4

Conclusion

01 A homologous TiO_2/NH_2-MIL-125(Ti) (LP-3) heterojunction is synthesized by a facile method to enhance the photocatalytic activity in the visible-light-induced aerobic oxidation of benzyl alcohol and degradation of tetracycline.

02 Benefit from The Ti^{4+}-Ti^{3+} intervalence cycles both in NH_2-MIL-125(Ti) and TiO_2, the LP-3 not only exhibits efficient transfer of photo-induced electron/hole pair, but also preserves the Ti^{3+} active species during the electron transfer process, resulting in high photocatalytic activity in the visible-light-induced aerobic oxidation of benzyl alcohol.

03 This work offers a facile strategy to preserve Ti^{3+} during the efficient transfer of photo-induced electrons through constructing a homologous TiO_2/NH_2-MIL-125(Ti) heterojunction.

24

Acknowledgement

- Group members at SCUT

Graduate Students:

Prof. Li Prof. Xiao Xiyi Li Yunhong Pi

- Funding supports
  - National Natural Science Foundation of China (21576093)
  - Guangdong Natural Science Funds for Distinguished Young Scholar
- The organizing committee of AIChE meeting

Thanks for your attention!

498981 Mid-Infrared Surface Plasmon-Enhanced Molecular Desorption

Abstract:

The strong light-matter interactions of localized surface plasmon resonances (LSPRs) offer opportunities to influence chemical reactions. Ultraviolet and visible LSPRs are known to impact reactions ranging from water splitting and hydrogen dissociation to ethylene epoxidation and methane formation. Doped semiconductor nanocrystals have emerged as a new class of materials that support LSPRs in the near and mid-infrared. Here, we use time-resolved *in situ* infrared spectroscopy to study the influence of mid-infrared LSPRs on molecular desorption. We find that the desorption rate for model molecules (e.g., indole, benzoic acid) on indium tin oxide (ITO) nanocrystals can be enhanced by as much as 2x upon illumination with mid-infrared light (~0.6-0.1 eV) at room temperature. The desorption rate is linearly dependent on light intensity, indicating a single photon process. Experiments as a function of LSPR energy, nanocrystal film thickness, and adsorbate concentration provide clues as to the mechanism of the enhancement. Our findings open new avenues to leverage low energy light for manipulating chemical reactions.

501053 Plasmonic Purification: Visible-Light-Driven Generation of Reactive Oxygen Species for Water Disinfection

Abstract:

Increasing access to clean drinking water is a global sustainability goal[1]. Solar energy can disinfect water through the generation of reactive oxygen species (ROS), but current methods utilize ultraviolet (UV) light, which makes up only 5% of the solar spectrum. Visible-light-driven ROS generation is a promising alternative but stable, highly absorbing materials with the proper oxidation and reduction potentials are necessary.

Here we report on the fabrication and characterization of a plasmonic photocatalyst for visible-light-driven ROS generation. Our plasmonic photocatalyst is a broadband absorber capturing 90 % of the solar spectrum centered around 600 nm. We test its ability to generate ROS in a custom-built continuous flow microreactor allowing rapid real-time measurements of H_2O_2 generation. Under UV filtered AM1.5G illumination our photo-microreactor generates ~5 Î¼M H_2O_2 per minute. We will discuss the performance and stability of our plasmonic photocatalyst and propose a reaction pathway for visible-light-driven ROS generation. The effects of photocatalyst geometry, residence time in the photo-microreactor and excitation spectrum will also be discussed.

1) World Health Organization. 2012. *Progress on Drinking Water and Sanitation.* Geneva, Switzerland: WHO Press.

499870 Efficient Photoreduction of Bicarbonate to Formate Catalyzed By Gold-TiO_2 composite Nanocatalyst Under Solar Light

Abstract:

TiO_2-catalyzed photoreduction of bicarbonate to value-added formate is studied in the presence of plasmonic gold nanoparticles. The effect of two positive hole scavengers, petroleum derived 2-propanol and green organic solvent glycerol, on the productivity of formate is investigated as well. Under solar light (solar simulator, AM 1.5 filter) the productivity with TiO_2 is 36.7 mM formate/g cat-hr in the presence of glycerol, which is significantly higher than 2-propanol (2.4 mM/hr.gm.catalyst). When illuminated with similar intensity 365 nm light with the photon energy high enough to excite valence band to conduction band electronic transitions in the TiO_2, the productivity of TiO_2 dramatically improves. A remarkably enhanced yield of formate (270 mM formate/g cat-hr) is achieved under solar light using gold-TiO_2 composite in the presence of glycerol as a hole scavenger. In contrast, no effect of gold nanoparticles was observed under 365 light. The improvement in product yield under visible light is attributed to the plasmon resonance induced hot electron transfer from gold nanoparticles to TiO_2. In addition, we studied the Au nanoparticle/TiO_2-catalyzed photoreduction of resazurin and photooxidation of Rhodamine B in parallel. As expected the addition of gold with TiO_2 improves the photoreduction of resazurin to resorufin under solar light. Surprisingly, gold nanoparticles adversely affected the photo-degradation rate of rhodamine B under both solar light and 365 nm light. This may be because of hydroxyl radical scavenging by gold nanoparticles.

483517 Solution Combustion Synthesis and Photoelectrochemistry of $Ga_xZn_{1-X}O_yN_{1-Y}$Â

Abstract:

Utilization of solar energy is a necessary path forward to deal with the looming energy crisis and worldwide effects of fossil fuel usage. A key component of this utilization is how to store excess solar energy for use on demand. One such method is splitting water into hydrogen and oxygen, both of which can be stored and either burned as fuel or used as feedstock components in industrial processes to synthesize high-value materials. Metal oxynitrides have shown great potential as photocatalysts for overall water splitting applications. Incorporation of nitrogen into metal oxides narrows the band gap to enable absorption of light in the visible spectrum. The typical synthesis of a metal oxynitride, however, requires annealing temperatures at or above 1000 °C and nitridation by flowing pure ammonia gas for anywhere from 1 to 48 hours. Such harsh synthesis conditions can limit the discovery of new candidate materials.

We have prepared an array of gallium zinc oxynitrides ($Ga_xZn_{1-x}O_yN_{1-y}$) as a model system using a combustion synthesis method. Gallium zinc oxynitrides can be synthesized in 30 minutes at 500 °C using this method. X-ray photoelectron spectroscopic analysis reveals that as much as 50% nitrogen substitution in the metal oxide lattice can be achieved, dependent on the ratio of gallium and zinc nitrates used as precursors. XPS also reveals that the nitrogen valency is that of a nitride material, as would be expected for a substitution into the crystal lattice. UV-visible spectroscopy also demonstrates that the band gap of the $Ga_xZn_{1-x}O_yN_{1-y}$materials can be controlled simply by altering the precursor ratio. X-ray diffraction measurements reveal shifts in the crystalline peaks from ZnO toward GaN, confirming the incorporation of nitrogen into the lattice. Of note is that the amount of gallium in the final product is lower than expected, based on the initial precursor mixture. This suggests that at least some of the gallium vaporizes during the reaction.

The as-synthesized products are capable of generating photocurrent from visible light irradiation and are able to carry out water splitting in pure water without any co-catalysts, such as Pt, IrO_2, or Rh/Cr_2O_3. Ongoing work is focused on improving product yield from the reaction, elucidating the effect of the reaction atmosphere on the final product, and maximizing photocurrent yield.

497211 Mixed Oxide Based Redox Catalysts for Hydrogen and Liquid Fuel Co-Generation Via a Hybrid Solar-Redox Scheme

Abstract:

Hydrogen is an attractive fuel due to its zero-emission combustion and high weight-based energy density. Among the various approaches for sustainable hydrogen generation, solar thermochemical water-splitting based on redox cycles of metal oxides represents an attractive approach. A key challenge for this process is the high temperature required for thermal decomposition of the metal oxide (>1200 °C). In addition, the need to balance the oxygen release and water-splitting properties of metal oxides often leads to low steam conversion. The current study investigates a hybrid solar-redox process for co-generation of hydrogen and liquid fuels using methane and solar energy at significantly lower reaction temperatures and an order of magnitude higher steam conversions. In such a process, a reduced metal-oxide-based redox catalyst is used to split water, producing concentrated hydrogen. The resulting (oxidized) redox catalyst is subsequently used to partially oxidize methane, forming syngas for Fischer-Tropsch synthesis. A rationalized strategy to optimize transition metal oxide-based redox catalysts for the scheme is proposed and validated. While monometallic transition metal oxides do not possess desirable properties, Density Functional Theory (DFT) calculations indicate that redox properties of mixed metal oxides such as equilibrium P_{O2} and vacancy concentration relationships can be adjusted via doping of heteroatoms. Based on such principles, a rationally developed, Ba and Fe-containing perovskite based redox catalyst was experimentally found to be able to achieve 90% steam-to-hydrogen conversion during water-splitting and over 90% syngas yield in the methane partial oxidation step under repeated redox cycles. Results from these experimental studies are then used in an ASPEN Plus® simulation model to evaluate the performance of the proposed hybrid solar-redox scheme

487941 Controllable Hydrothermal Conversion from Ni-Co-Mn Carbonate Nanoparticles to Spheres

Abstract:

Abstract: Starting from Ni-Co-Mn carbonate nanoparticles prepared by microreaction technology, uniform spherical particles of $Ni_{1/3}Co_{1/3}Mn_{1/3}CO_3$ with size of 3-4 Î¼m were obtained by a controllable hydrothermal conversion with the addition of $(NH_4)_2CO_3$. Based on characterizations on evolution of morphology and composition with hydrothermal treatment time, we clarified the mechanism of this novel method as a dissolution-recrystallization process as well as the effects of $(NH_4)_2CO_3$ concentration on the morphology and composition of particles. By changing concentrations and ratio of staring materials for nano-precipitation preparation, we achieved monotonic regulation on the size of the spherical particles, and the synthesis of $Ni_{0.4}Co_{0.2}Mn_{0.4}CO_3$ and $Ni_{0.5}Co_{0.2}Mn_{0.3}CO_3$, respectively. In addition, the spherical particles with core-shell structure were preliminarily verified to be available by introducing nano-precipitates with different compositions in the hydrothermal treatment in sequence.

492510 Oxidative Chemical Vapor Deposition of Polyaniline: Influence of Process Conditions on Film Chemistry and Electrochemical Performance

Abstract:

The use of conducting polymers as the active material for supercapacitors is a promising approach to improve the performance of supercapacitors. Conducting polymers enhance charge storage capacity through Faradaic redox reactions, and have the ability to be p- or n-doped. They are typically deposited via solvent-based techniques such as chemical bath deposition, electrodeposition, and casting from suspension. Nevertheless, challenges exist with making ultrathin (5-10 nm) uniform conformal coatings via wet chemistry routes, especially with poor solubility of many conducting polymers and poor accessibility of the highly tortuous pore space within nanostructured electrodes. To conveniently bypass these challenges, polyaniline (PANI) is synthesized via oxidative chemical vapor deposition (oCVD) in a single-step, solvent-free process using aniline monomer and antimony pentachloride oxidant, to deposit ultrathin films into Mo_2C carbide-derived-carbon (CDC) electrodes (Figure 1).[1] we are realizing a strategy proposed by Simon and Gogosti[2] to improve both the energy and power densities of electrochemical capacitors by integrating conducting polymers into nanostructures electrodes.

By understanding the oCVD PANI synthesis parameters (reagent flow rates, reactor pressure, substrate temperature), the reaction and diffusion rates can be carefully controlled to optimally coat the porous carbide derived carbon (CDC) electrode. The oxidant flowrate, substrate temperature, and reactor pressure were varied, and their influence on film chemistry and supercapacitor performance was explored. The study reveals that a higher substrate temperature, pressure, and oxidant flowrate are critical for depositing emeraldine PANI for optimal electrochemical performance. Spectroscopic results,[3] such as FTIR, SEM and XPS, indicate that a substrate temperature of 90 °C is needed to minimize the formation of oligomers during polymerization. Increasing the oxidant flowrate to nearly match the monomer flowrate favors the deposition of PANI in the emeraldine state. Lower substrate temperatures (Low-T), such as 25 °C, lower pressures (Low-P) and lower oxidant flow (Low-O) generally lead to poorer quality PANI films that results in more oligomers. By optimizing the oCVD processing conditions, CDC electrodes integrated with oCVD PANI exhibits more than double the gravimetric capacitance (115 F/g) vs bare CDC electrodes (52 F/g) and a 79% capacity retention after over 10,000 cycles.[4] Interestingly, the optimally performing PANI-CDC devices have a porous PANI morphology as determined by SEM, which may facilitate ion transport, improve scan rate performance, and impart electric double layer capacitance in addition to the intrinsic faradaic pseudocapacitance.

1. Smolin, Y. Y.; Van Aken, K. L.; Boota, M.; Soroush, M.; Gogotsi, Y.; Lau, K. K. S., Engineering Ultrathin Polyaniline in Micro/Mesoporous Carbon Supercapacitor Electrodes Using Oxidative Chemical Vapor Deposition. *Adv. Mater. Interfaces* **2017**, DOI: 10.1002/admi.201601201

2. Simon, P.; Gogotsi, Y., Materials for Electrochemical Capacitors. *Nat. Mater.* **2008**, *7*, 845-854.

3. Smolin, Y.Y.; Soroush, M.; Lau, K. K. S., Oxidative Chemical Vapor Deposition of Polyaniline Thin Films. *Beilstein J. Nanotech.* **2017** (under review).

4. Smolin, Y. Y.; Soroush, M.; Lau K. K. S., Influence of oCVD Polayanilnie Film Chemistry in Carbo-Based Supercapacitors. *Ind. Eng. Chem. Res.* **2017** (under review).

Figure 1. Left: oCVD process where the oxidant ($SbCl_5$, orange) and monomer (aniline, green) vapors enter the reaction chamber and surface polymerize onto the porous electrode. Right: A comparison of the cyclic voltammograms and cycle life for both bare (black) CDC and oCVD PANI-CDC (colored) at various oCVD process conditions (at 20 mV/s).

497915 High Performance Pillared V_2O_5 and MnO_2 Cathodes for Lithium Ion Batteries

Abstract:

Lithium ion batteries (LIBs) are used extensively in portable electronics because of their high volumetric energy densities. Beyond this, LIBs are becoming more and more popular for aerospace and electric vehicle applications [1-3]. Layered lithium metal oxides, such as $LiCoO_2$, are commonly used as the cathodes given their high capacities and energy densities. However, the poor mechanical and thermal properties of these materials restrict their application under some challenging use conditions. For example, in military vehicle applications, the battery will be exposed to a wide operating temperature range. This is very challenging for $LiCoO_2$, as oxygen liberation will happen at ~200°C [4]. Also, when LIBs are charged and discharged, repeated lithium ion insertion and extraction leads to mechanical fatigue and deformation of the lattice structure [5], which causes capacity fade and increased thermal runaway hazards.

To address challenges associated with the layered oxide materials and improve their thermal and structural stabilities, as well as their rate capabilities and cyclabilities, modifications of the nanoscale structure have been carried out. Our approach is to intercalate nanoscale pillars between the layers to provide support [6-7], thus leading to a more robust structure. Vanadium pentoxide (V_2O_5) is a promising cathode material for LIBs, however it suffers from poor structural stability and limited rate capability [8]. Our research demonstrates significant improvements in the thermal and electrochemical properties of V_2O_5 xerogels via nanopillaring with Al_{13} Keggin ions.

X-ray diffraction (XRD) and transmission electron microscopy (SEM) showed that the interlayer spacing of V_2O_5 xerogels (V_2O_5G) increased ~15% through intercalation of Al_{13} Keggin ions, which can provide a more favorable pathway for Li^+ diffusion. Aluminum contents as high as 9 wt% were achieved. In addition, thermal gravimetric analysis (TGA) illustrated that the pillared materials possessed better thermal stability. The V_2O_5-Al_{13} was stable up to 400 °C compared to 350 °C for V_2O_5G.

Fundamental electrochemical properties of the V_2O_5G and V_2O_5-Al_{13} were characterized using cyclic voltammetry. V_2O_5-Al_{13} displayed improved redox activity supported by a prominent increase in current density relative to the untreated gel. To characterize the cycling performance, rate capability tests were conducted to determine capacity retention and capacities at high rate for V_2O_5G and V_2O_5-Al_{13}. V_2O_5G could only retain 36% of the initial capacity when switching from low rate (C/10) to high rate (C/2) discharge, while V_2O_5-Al_{13} was able to retain 60% of the initial capacity. When returning to low rate (C/10), V_2O_5-Al_{13} also showed better performance holding 80~95% of its original capacity. V_2O_5G, however, leveled off at 60%. As long-term stability at high rates of discharge is a challenge for LIBs [9], cycling at C/2 was also carried out to investigate their long-term cyclability. The results clearly show that V_2O_5G faded quickly from 75 to 52 mAh/g (based on active material) during its first 20 cycles, while V_2O_5-Al_{13} started at a much higher capacity of 145 mAh/g and maintained 86% of this capacity after 50 cycles. The V_2O_5-Al_{13} material also maintained 75% of its initial capacity after 100 cycles, demonstrating much better performance than V_2O_5G.

Given the positive results achieved by nanostructuring V_2O_5G, we also pillared manganese dioxide (MnO_2). Because of its low cost, non-toxicity, environmental friendliness and relative abundance [11-13], MnO_2 is particularly attractive for use as a cathode material. Two morphologies of layered MnO_2 have been successfully synthesized and characterization methods were used to defined their structural and compositional properties. The first discharge capacity for the Al_{13} Keggin ion pillared MnO_2 (MnO-Al_{13}) was up to 265 mAh/g. These and other results from pillaring oxides will be presented in this paper.

[1] Dunn, B., Kamath, H., Tarascon, J. M. Science **2011**, 334(6058), 928-935.

[2] Armand, M., Tarascon, J. M. Nature **2008**, 451(7179), 652-657.

[3] Hayner, C. M., Zhao, X., Kung, H. H. Annual review of chemical and biomolecular engineering **2012**, 3, 445-471.

[4] Baba, Y., Okada, S., Yamaki, J. I. Solid State Ionics **2002**, 148(3), 311-316.

[5] Vetter, J., Novák, P., Wagner, M. R., Veit, C., Möller, K. C., Besenhard, J. O., Hammouche, A. Journal of power sources **2005**, 147(1), 269-281.

[6] Petridis, D., Kaviratna, P. D. S., Pinnavaia, T. J. Journal of Electroanalytical Chemistry **1996**, 410(1), 93-99.

[7] Wang, L., Sakai, N., Ebina, Y., Takada, K., Sasaki, T. Chemistry of materials **2005**, 17(6), 1352-1357.

[8] Whittingham, M. S. Chemical reviews **2004**, 104(10), 4271-4302.

[9] Etacheri, V., Marom, R., Elazari, R., Salitra, G., Aurbach, D. Energy & Environmental Science **2011**, 4(9), 3243-3262.

[10] Du, G., Wang, J., Guo, Z., Chen, Z., Liu, H. Materials Letters **2011**, 65(9), 1319-1322.

[11] Morales, J., Sanchez, L., Bach, S., Pereira-Ramos, J. P. Materials Letters **2002**, 56(5), 653-659.

[12] Komaba, S., Kumagai, N., Chiba, S. Electrochimica Acta **2000**, 46(1), 31-37.

498591 2D Transition Metal Carbides (MXenes): Synthesis and Applications in Electrochemical Energy Storage

Abstract:

Two-dimensional (2D) materials, such as graphene, metal oxides/hydroxides, dichalcogenides, hexagonal boron nitride, and others are garnering increasing attention due to their unique properties. About 4 years ago, our group discovered a new class of 2D transition metal carbides and nitrides we labeled MXenes, because they are obtained by selectively etching the A-layers from their three-dimensional (3D) layered, parent compounds the $M_{n+1}AX_n$, or MAX, phases. So far, synthesis of ~20 MXenes with different chemistries have been achieved. MXenes offer a unique combination of metallic conductivity or semiconducting properties and hydrophilicity, coupled with good mechanical properties. They have shown superior capacity for reversible intercalation of many metal cations (e.g. Li^+, Na^+, K^+, Mg^{2+}, Al^{3+}, etc.), which, in turn, has led to their use as electrode materials in energy storage devices, such as supercapacitors, Li-ion and Na-ion batteries/capacitors, and Li-S batteries. Moreover, several methods, including introduction of interlayer spacers, porosity, and hybridization with metal oxides or conductive polymers, have been utilized to significantly improve the electrochemical performance of MXenes. As a new family of 2D materials, MXenes also show great potential in the application of fuel cells, transparent conductors, separators, and catalysis, etc.

498472 Design, Synthesis, and Characterization of Mixed Ionic/Electronic Conducting Surface Layers Adsorbed on Metal Oxide Particles

Abstract:

Inexpensive grid-scale storage remains a major hurdle slowing the wide-scale adoption of renewable energy by public power utilities. Flow batteries have been proposed as one viable solution. Whereas, the capacity of a traditional battery is limited by its packaging, flow batteries pump redox active fluids from external storage tanks through a flow cell where electrical energy is extracted or stored via reversible redox reactions. In this way, the capacity of the battery becomes scalable independent of the specific redox chemistry and battery geometry. Semi-Solid flow batteries (SSFBs) use â€œflowable electrodesâ€ for both the anode and cathode materials that consist of a mixture of lithium containing metal oxide particles and carbon black particles. Carbon black is added to these dispersions to improve the utilization of lithium as the particles flow by the electrodes. Unfortunately, the presence of this percolated carbon black network leads to significant increases in viscosity, both at low and high shear rates. In this work, we outline the synthesis of a composite polyelectrolyte/conjugated polymer adsorbed surface layer on cationic silica particles. This layer is formed through the adsorption of poly(styrene sulfonate) onto oppositely charged metal oxide particles and subsequent polymerization of the surface polyelectrolyte with 3,4-ethylenedioxithiophene (EDOT) monomer. The PEDOT:PSS layer is characterized as a function of EDOT:SO_4^{-2} loading using a combination of light scattering, neutron scattering and dielectric spectroscopy to understand the connection between the surface layerâ€™s composition, nanostructure and electrical properties. These properties are then connected to the rheological performance of suspensions in propylene carbonate and the mechanical and electrical properties identified.

502705 Conductive Membrane Coatings for Improving Current Density in Redox Flow Batteries

Abstract:

Global electricity production from renewable sources has grown dramatically in recent years. Some of these forms of electricity production are intermittent in nature, and have proven difficult to incorporate into regional electrical grids. Grid-scale electricity storage methods are a promising solution to the inefficiencies of unpredictable electricity generation. Redox flow batteries have demonstrated the ability to be used as a reasonably low-cost, long-term, grid-scale electrical storage method. Specifically, vanadium redox flow batteries (VRBs) have been of special interest due to their chemical stability, long life cyclability, and potential for high electrical capacity compared to other redox chemistries. Accompanying these advantages are some research challenges. There have been a number of efforts to improve the electrical conductivity of VRBs by physically or chemically altering the electrodes of the cell. Most VRBs use a carbon or graphite felt as porous electrodes in the battery â€" the large surface area from the felt is used to increase the number of reaction sites at which the redox reaction can occur. Many modifications to the electrodes have been investigated, including using various highly conductive carbons, metal catalysts, and covering the carbon felt with hydrophilic surface groups. Less attention has been spent improving the membrane surface properties to alter the electrical and ionic conductivity in the cell. A new approach to improve the rate capability of VRBs involves improving the electrical conductivity of the surface of the membrane itself using air-assisted electrospray deposition of conductive carbons.

A conductive coating of carbon nanotubes (CNTs) and Nafion dispersion in water was used to coat a Nafion 117 membrane via air-assisted electrospraying to improve the rate capability and cycling performance of VRBs. It was found that electrospraying a highly conductive coating directly onto the surface of the membrane allowed for stable cycling performance at nearly double the current density that was afforded by the pristine Nafion membrane. A templating technique was used during the electrospraying process to allow for alternating domains of coated and uncoated membrane surface, which helped reduce the restriction of proton transport across the membrane, further improving rate capability and capacity retention of the VRB. The interfacial resistance between the membrane and the electrode was greatly diminished with the addition of a very small mass (<0.01 mg/cm^2) of CNTs. This method has shown to be a fast, simple, and scalable technique for improving the rate capability of vanadium redox flow batteries.

This technique was subsequently extended to a quinone-based aqueous redox flow battery system. While the results were more modest when compared to the VRB system, improvements in current density were still seen with the use of a membrane with a conductive coating.

492522 Impact of Ionomer Resistance in Nanofiber-Nanoparticle Electrodes for Ultra-Low Platinum Fuel Cells

Abstract:

Recently, in our laboratory, a new simultaneous electrospinning and electrospraying (E/E) process was developed to produce unique nanofiber-nanoparticle electrodes for polymer electrolyte membrane (PEM) fuel cells that resulted in high power densities at ultra-low platinum (Pt) loadings. These results demonstrated the impact of electrode (catalyst layer) morphology on fuel cell performance. Although promising, these previous results did not explore the impact of all of the electrode morphology parameters (*e.g.*, fiber/catalyst content, fiber size, fiber/catalyst structure, porosity, *etc.*) on power density (kW/cm^2), Pt loading (mg/cm^2), and subsequently Pt utilization (kW/mg). In this study, E/E electrodes were systematically fabricated to fundamentally investigate the impact of the ionomer (Nafion) in the catalyst layer on transport resistances. To our knowledge, previous studies have not been able to independently study mass transfer resistance, proton transfer resistance, and porosity of the catalyst layer on fuel cell performance, because of their coupled effects within the catalyst layer. Utilizing the E/E process, the effect of the ionomer in the electrodes on mass transfer and proton transfer resistances can be investigated separately, which will significantly impact the design of fuel cell catalyst layers and elucidate various transport phenomena effects in the catalyst layer for ultra-low Pt loadings. Overall, the results show an increase in fuel cell power density with increasing ionomer content in nanofiber-nanoparticle electrodes until a specific ionomer content is reached. After this optimal ionomer content, the fuel cell power density decreases with increasing ionomer content. Furthermore, results show that at zero to low ionomer content, there is insignificant difference in power density, yet the catalyst layer resistance decreases due to the increase of triple phase boundary in the catalyst layer, *i.e.,* decrease in proton transfer resistance. As the ionomer content increases further, the catalyst layer resistance begins to increase because of the oxygen mass transfer resistance due to the ionomer thin film resistance. Therefore, there is an optimum ionomer content to minimize proton transfer resistance and oxygen mass transfer resistance. Furthermore, the oxygen mass transfer resistance seems to have more of an impact on the power density than the proton transfer resistance.

499523 Plant Cell Wall Inspired Nanoscale Materials for Renewable Energy Applications

Abstract:

Nature has always been a source of inspiration and knowledge for researchers to gain insight into material properties at nanoscale and design next generation materials. Many naturally abundant molecules have the potential to be converted to valuable products. Such materials can be environment friendly and cheaper alternatives of artificially synthesized molecule based systems. Current initiatives to reduce the impact of global warming and climate changes demand best utilization of renewable energy sources. If, in addition to using renewable energy, we can use renewable polymers to design efficient and durable ion transporting materials for energy conversion systems, it will protect the environment. However, the natural materials have to be highly functional, durable and cheap so that they can compete with artificially synthesized, relatively expensive materials. The current work takes inspiration from plant cell walls and utilizes the major polymers (Cellulose, Lignin) present in plant cell walls to design ion conducting materials that can be used for renewable energy propelled applications, such as water electrolysis cells, fuel cells and supercapacitors. These applications require nanoscale thick materials at the electrode interface for efficient ion conduction and redox reactions to occur at the electrodes. The current work focuses on exploring how cellulose and lignin based ionomers behave under varied confined environments in several hundreds of nm thick films. The ion conduction and optical properties are measured to explore local hydration environment and find the correlation among hydration level, film composition and ion transport properties in such nanoscale materials using fluorescent dyes. These structure-property relationships will be a key to optimize the ion conductivity and mechanical strength. By understanding such complex bioinspired systems, we can gradually move towards presenting a natural, green alternative of synthetic ion conducting polymers for renewable energy applications.

501482 Engineering High Performance and Durable PGM-Free Electro-Catalysts for Oxygen Evolution Reaction in PEM Water Electrolysis

Abstract:

The rising concerns related to rapid depletion of conventional fossil fuels and alarming indications of global warming has triggered the pursuit of environment friendly and sustainable alternative energy sources. In this regards, hydrogen (H_2) gas having non-carbonaceous nature and higher energy density than carbonaceous energy sources is indeed the foremost energy carrier[1]. The generation of clean and sustainable hydrogen via acid assisted i.e. proton exchange membrane (PEM) based water electrolysis is hitherto hailed as one of the most efficient and promising technologies among all other conventional hydrogen production methods[2]. Although PEM water electrolyzers offer a reliable way to produce high-purity hydrogen in large scales, commercial development of this technology has been largely restricted due to the need for costly and scarce platinum group metal (PGM) based electro-catalysts such as Pt, RuO_2, IrO_2; which demonstrate superior electro-catalytic activity towards energy intensive and sluggish oxygen evolution reaction (OER) in PEM water electrolysis[2]. Therefore, to overcome this bottleneck, exploration of earth abundant, economic and PGM-free electro-catalysts, exhibiting comparable or superior electro-catalytic activity and robust durability compared to state-of-the art noble metals based electro-catalysts in harsh acidic media is extremely desired, which can lead to significant reduction in capital cost of the PEM water splitting process[3].

In line with this objective, exploiting theoretical first principles approaches, we have engineered anionic fluorine (F) doped transition metals non-oxide pnictide (TMN) based electro-catalyst for OER in PEM water electrolysis. The dopant-F plays a vital role in enhancing the electro-catalytic activity by altering the electronic structure of the electro-catalyst. The as-synthesized electro-catalyst with F content of 10 wt. % consequently exhibits excellent electro-catalytic performance, outperforming the benchmark IrO_2 for OER in PEM water splitting. The as-synthesized electro-catalyst powder was coated on porous titanium (Ti) substrate (total catalyst loading=1 mg/cm^2) and used as an anode. Electrochemical characterization of the electro-catalyst has been carried out in three-electrode configuration system, using 1N sulfuric acid (H_2SO_4) solution as the proton source as well as the electrolyte. Pt wire and Hg/Hg_2SO_4 are used as a counter electrode and reference electrode (+0.65 V with respect to the normal hydrogen electrode, NHE) respectively. Electrochemical evaluation has been conducted at a scan rate of 10 mV/sec and at temperature of 40°C. The generated PGM-free electro-catalyst exhibits an onset potential of ~1.43 *vs* NHE, similar to that of IrO_2 but significantly lower charge transfer resistance (R_{ct}) than benchmark IrO_2. In addition, the as-synthesized electro-catalyst with F content of 10 wt. % displays a remarkable ~1.4 fold higher electro-catalytic OER activity (i.e. current density at 1.5 V) than that of IrO_2 approaching a current density of ~ 10 mA/cm^2 at an overpotential of ~ 245mV. Chronoamperometry tests conducted in 1N H_2SO_4 solution at ~1.5 V (*vs* NHE) for 24 hours also show marginal loss in current density, indicating good electrochemical stability of the as-prepared electro-catalyst. In summary, we have synthesized highly efficient PGM-free F doped TMN pnictide based electro-catalyst for OER in PEM based water electrolysis. The superior electrocatalytic activity of this electro-catalyst is attributed to the modification of the electronic structure (as evidenced by the theoretical studies) and lower charge transfer resistance (i.e. lower activation polarization). Thus, the present electro-catalyst system is indeed promising and beneficial for the economical and efficient hydrogen production from PEM water splitting. Results of these studies will be presented and discussed.

References:

[1] A. S. Aricò, V. Baglio, N. Briguglio, G. Maggio, S. Siracusano, *Fuel Cells: Data, Facts and Figures* **2016**.

[2] S. D. Ghadge, P. P. Patel, M. K. Datta, O. I. Velikokhatnyi, R. Kuruba, P. M. Shanthi, P. N. Kumta, *RSC Advances* **2017**, *7*, 17311-17324.

[3] S. A. Grigoriev, V. I. Porembsky, V. N. Fateev, *International Journal of Hydrogen Energy* **2006**, *31*, 171-175.

502728 Electroactivity and Stability Analysis of Nickel Oxide Nanoclusters Deposited on Graphene with Ball Milling and Microwave-Assisted Deposition for Glucose Sensing and Fuel Cells

Abstract:

In order to realize commercial sensors and fuel cells, stable electroactivity is a major concern for fuel cell and sensor development. Commercial graphene was used as substrate to form nickel oxide nanoparticles from nickel salts by ball milling and microwave assisted deposition to form 10wt% NiO catalyst. Nanoclusters are characterized by SEM and XRD to identify size and crystallinity. Cyclic voltammetry and amperometry are used to assess electrocatalytic activity in 0.05 M NaOH with and without glucose. A conventional three electrode cell was used with catalyst ink doped GCE working electrode, AgCl reference electrode, and Pt wire counter electrode. Ball milled catalyst samples versus ball milled and microwave assisted catalyst samples are compared directly on the basis of open circuit potential, peak current density, sensor sensitivity and limit of detection (LOD), and cycling stability. It is demonstrated that NiO cheaply catalyzes the glucose oxidation reaction and that microwave assisted samples show longer stability during cycling.

499781 A Combined High-Throughput Computing and Machine Learning Study Reveals Hydrogen Storage Performance Ceilings of Metal-Organic Frameworks

Abstract:

MOFs with high hydrogen storage capacities are revealed via high-throughput screening and machine learning on nearly half a million compounds. Screening based on grand canonical Monte Carlo and other techniques identified over 2,000 MOFs that exceed the hydrogen capacity of the benchmark compound MOF-5 by more than 15%. Our data reveals the existence of a celling in usable volumetric capacity near 40 g H_2/L, indicating that new design directions for improved volumetric performance are needed. Principal component analysis (PCA) and decision tree (DT) classification were performed on our database to inform design rules. PCA reveals the geometrical properties of MOFs (e.g., surface area, pore volume, pore diameter, density, and void fraction) that correlate with high hydrogen storage capacities. Furthermore, decision tree classification suggested design directions by delineating geometrical properties with acceptable ranges. These data provide guidelines for the design of MOFs with much improved volumetric performance.

482163 Designing Light Propelling Nanomotors

Abstract:

Introduction

It has attracted much interest to design a self-powered artificial nanorobotic system which mimic the behavior of the motile bacteria due to both scientific merits as well as its potential applications. Previously, it has been demonstrated that asymmetric redox reaction on bi-metallic nanowire can produce the electric field and propel itself in solution. This autonomous motion shows that artificial inorganic nanomaterial can be used as nanomotor which harvests energy from the environment.

Here, we will present a novel approach to rational designed micro-swimmer which harvests energy from absorbed photons by photoeletrochemical (PEC) reaction. Upon illumination, the photoelectrochemical reaction on the swimmer surface generates the asymmetric ionic enviroments required to propel the migration of micro-swimmer.

We will present seveal approaches including asymmetric Janus nanostructure and asummetric photonic structure. In this design, we focus on controllability and programmability of the micro-swimmer. The Individual Janus nanotree is designed that it can sense the direction of light vector and direct to illumination direction. Or the nanomotor can be spectrally engineered to response to specific wavelength of light. As the light with different frequencies can be applied to nanomotor, it offers great flexibility and controllablity to nanomotor/nanorobot design.

Figure a. Schematic of micro-swimmer. **b.** The trajectory of a pristine nanotree spells "nano" as navigated by light. **c.** Sequential images of the green algae, *E. Gracilis, pristine microswimmer, CSPTMS treated micro-swimmer* suspension in aqueous solution with illumination from the right side

Reference:

B. Dai, J. Wang , Z. Xiong, X. Zhan, W. Dai, C.C. Li, S. Feng, J. Tang, Programmable Artificial Phototactic Microswimmer, *Nat. Nanotechnol. 11, 1087–1092(2016)*

486462 Improving Electron Transport in TiO_2 Electrode Via Nanostructure Engineering

Abstract:

Titanium dioxide (TiO_2) is one of the most widely used semiconductors in photovoltaics and photocatalysis because it is nontoxic, abundant, stable and photoactive. However, the wide bandgap, low electron mobility and short minority carrier diffusion length of TiO_2 limit its quantum efficiency in these applications. In this work, we present a solution chemical approach for making TiO_2 nanostructures for improving the electron transport in nanostructured TiO_2 electrodes.

490844 Silicon-Based Infrared Photodetectors Enabled By Hot Electrons

Abstract:

Conventionally, photonic infrared (IR) detectors employ low band gap materials such as InGaAs, InSb, or HgCdTe. However, these materials include elements that are rare, expensive, or toxic. Past research indicates that crystalline Si (c-Si), which is a much cheaper and more abundant element, could be used for IR detection when metal electrodes are cleverly nanostructured. In this type of photodetection systems, the IR light with energies below the c-Si band gap is strongly absorbed by the metal structures, rather than by c-Si. The photoexcited electrons in the metal can then be injected into the conduction band of c-Si before being thermalized and electric current can be generated. These non-thermalized electrons, called hot electrons, enable the detection of IR light with energies below the c-Si band gap. For efficient transport of electrons in the metal before thermalization, the metal layer should be as thin as approximately the electron mean free path. Accordingly, the metal layer thickness should be only a few tens of nanometers. To induce strong optical absorption in such a thin metal layer, surface plasmon polaritons (SPPs) can be excited at the metal surface. Previous studies on hot electron photodetection utilized small-scale metamaterials or deep trench resonators to have strong resonant absorption of SPPs in thin metal films on c-Si at the desired frequencies. However, these structures should be fabricated with a high precision because the metal structure sizes determine resonances. Accordingly, in many cases, expensive techniques such as electron beam lithography have been commonly used to fabricate the structures. However, for mass production, it is important to obtain metallic structures that do not require expensive techniques and tolerate practical fabrication errors. In this study, we use metal metasurfaces that can be fabricated by scalable, inexpensive techniques and achieve a broad-band IR absorption of over 95% in 15-nm-thick metal films. This unprecedented strong absorption, in terms of both the absorptance magnitude and the band width, is enabled by a new scheme where the light takes multiple passes within the c-Si substrate. During the passage, light is preferentially absorbed by the thin metal layer that is on one side of the substrate. Absorption on the other side is efficiently eliminated by using a dielectric layer. In our effort, the surface of the c-Si substrate where thin meal film is deposited is structured by a simple optical lithography. The structured surface admits the incident light into the substrate and prevents the light from leaking out of the substrate. In our scheme of multiple light passes, extremely strong resonances are not necessary and fabrication errors would not destroy the optical properties appreciably. In this talk, we will discuss the details of the optical absorption in our scheme. We will also present our experimental results on the electronic characteristics of our hot electron devices.

494052 Enhancing Light Absorption and Emission in Quantum Dot Solids Using Al Nanostructures

Abstract:

Nanocrystals such as CdSe/CdS core/shell quantum dots (QDs) are desirable optical materials with tunable band gaps, high quantum yields, and large Stokes shifts. These materials are typically synthesized in solution, but integration into optoelectronic devices requires the creation of nanocrystal solids. These solids must preserve the isolation between individual nanocrystals in order to maintain the benefits of quantum confinement, while simultaneously allowing for the transport of excitons or electrical currents. Nanocrystal solids exhibit distinctly different optical properties from their solution-phase counterparts, and complete engineering of these devices requires a detailed understanding of these changes in optical properties. Here we aim to manipulate the optical properties of QD solids by incorporating plasmonic nanostructures that enhance and direct light absorption as well as tune the local density of optical states, which influences light emission.

We study the influence of arrays of plasmonic nanostructures on the absorption and the emission from QD solids. The arrays consist of aluminum disks and rings, which both scatter incident light and possess strongly enhanced local electromagnetic fields. To avoid quenching and to provide a uniform surface for the deposition of the QD solid, the plasmonic nanostructures are completely embedded within a thin film of alumina. Aluminum is chosen over silver or gold by matching the plasmonic nanostructures to the absorption of the semiconductor nanocrystals. We study the absorption and the emission from both a continuous QD solid and from patterned QD solid regions.

Electromagnetic simulations are performed using finite-difference time-domain numerical simulation software to optimize the designs and understand the interactions between the Al nanostructures and the QD solids. The Al nanostructures are 30 nm tall and embedded in 34 nm of alumina to provide a thin barrier that prevents direct contact of the QD solid with the aluminum nanostructures. The optical properties of the QD solid are modeled using data from variable angle spectroscopic ellipsometry.

For a 60 nm thick QD film deposited on top of the alumina layer, we observe greater than a threefold increase in absorption within the continuous QD solid and greater than a ninefold increase in absorption within the patterned QD solid regions due to the presence of the embedded aluminum nanostructure array. Furthermore, we find that the magnitude of the scattering cross section of the plasmonic nanostructure contributes more to the overall absorption enhancement than the perfectly matched resonance of the individual aluminum nanostructure. This indicates the importance of the scattering properties of these embedded arrays to redistribute the incident light and increase the path length within the absorbing QD solid. The addition of a back reflector prevents the transmission of light through the system and increases the absorption within the QD solid. However for certain nonabsorbing spacer layer thicknesses between the nanostructure array and the back reflector, the reflected light destructively interferes and greatly reduces the electric field and thus the absorption within the QD solid.

In conjunction with absorption simulations, the emission from the QD solid is important to consider as the experimentally measurable luminescence consists of both the absorption of incident light and the emission from the individual QDs in the QD solid. Using a dipole to model the emission from the QDs, the emission is averaged over the entire periodic array to determine the emission enhancement from the embedded aluminum nanostructure array. Thus with these two simulations, the overall luminescence enhancement from the embedded nanostructured array is determined.

To support these simulation results, we fabricate aluminum nanostructured arrays embedded in alumina through a combination of electron beam lithography, thermal evaporation, and RF sputtering. We use a home-built fluorescence lifetime imaging microscopy setup with both time-resolved lifetime and steady state emission spectra capabilities to characterize the changes in optical properties due to the presence of the embedded aluminum nanostructured array. To create patterned QD solids, we require a ligand exchange from nonpolar ligands common during the synthesis process including oleate and myristate to polar ligands like 3-mercaptopropionic acid. In doing so, a blueshift in the emission wavelength from 588 to 584 nm is observed along with a significant drop in luminescence. A more drastic change in the fluorescence lifetime is observed from over 15 ns to less than 1 ns. These changes in the emission wavelength and the fluorescence lifetime are expected from the ligand exchange process due to the addition of surface defects and decrease in surface passivation of the QDs. This insight will help guide the efficient integration of plasmonic nanostructured arrays into optoelectronic devices.

495772 Precise Control over the Morphology and Dopant Distribution in Colloidal Metal Oxide Nanocrystals

Abstract:

Colloidal synthesis of doped metal oxide nanocrystals provides a great opportunity and easy route to generate materials that has unique optoelectronic properties with promising applications such as smart windows, displays, sensing and photo-catalysis etc. By introducing the free carriers with different type of dopants (n- or p-type) in the metal oxide nanocrystals, their surface plasmon resonance can be tuned precisely from near IR to mid-IR range. Similarly like metals, the optical response of plasmonic metal oxide nanocrystals can be manipulated by controlling the shape, size of the nanocrystal and free electron concentration. The effect of nanocrystal shape and size on the enhancement of their local electrical field strength and surface plasmon resonance have paved the way for new technologies and better sensing opportunities. The sharp faceted nanocrystals exhibit enhanced electric fields at corners and edges, which give us an opportunity to explore different morphologies of the NC for sensing application. Here, we will be presenting a solution route to synthesize plasmonic metal oxide nanocrystal (doped Indium Oxide) with defined shape, size and radial distribution of dopant in the nanocrystals. Also, with co-doping (cation, anion or both) in these nanocrystals, we can shift the surface plasmon resonance to higher energies and can also influence the shape of the nanocrystals. Further, we will present near field enhancement property of single nanocrystals via EELS mapping and quantify both near field and far field plasmon property via COMSOL electromagnetic simulations.

497411 Enhanced Optoelectronic Functionality of Photovoltaic 2D Crystal-Nanoantenna Hybrids

Abstract:

Enhanced optoelectronic functionality of photovoltaic 2D crystal-nanoantenna hybrids

D. Keith Roper[a,b*], Gregory T. Forcherio[c], Jeremy R. Dunklin[d], Mourad Benamara[e], Luigi Bonacina[f], Yana Vaynzof[g], Claudia Backes[h]

[a] *Microelectronics and Photonics Graduate Program, University of Arkansas, Fayetteville, AR 72701*

[b] *Department of Chemical Engineering, University of Arkansas, Fayetteville, AR 72701*

[c] *Army Research Laboratory, Adelphi, MD 20783*

[d] *National Renewable Energy Laboratory, Golden, CO 80401*

[e] *Institute for Nanoscience and Engineering, University of Arkansas, Fayetteville, AR 72701*

[f] *GAP-Biophotonics, Université de Genève, Genève, CH 1211*

[g] *Kirchhoff-Institute for Physics and Centre for Advanced Materials, Ruprecht-Karls University Heidelberg, 69120 Heidelberg, Germany*

[h] *Chair of Applied Physical Chemistry, Ruprecht-Karls University Heidelberg, 69120 Heidelberg, Germany*

Two-dimensional (2D) nanocrystals (NC) offer a direct bandgap spanning the solar spectrum as well as enhanced electron mobility and gate tunability. Decoration of 2DNC by nanoantenna (NA) has enhanced measured photocurrent. But challenges in simulating, characterizing and fabricating 2DNC-NA hybrid structures has constrained their implementation in functional photovoltaic devices.

This work examined optical modulation of photocurrent, carrier injection and second harmonic generation (SHG) in 2DNC, i.e., graphene and transition metal dichalcogenide (TMD), via optically active NA. Monolayer 2DNC used were chemical vapor deposited (graphene) or liquid exfoliated from bulk (TMD). 2DNC were decorated by gold (Au) and silver (Ag) NA through three methods (i) evaporation, (ii) drop-casting and (iii) direct reduction (Fig 1; [2]). Electron energy loss spectroscopy (EELS) was used to predict and induce plasmon bright, dark, and hybrid modes. EELS enabled quantitative femtosecond-scale measurement of spectroscopic plasmon dephasing and nanometer-resolved mapping of electric fields on the 2DNC-NA hybrids (Fig 2). Coupled and discrete dipole simulations were used to characterize radiative and intraband dephasing in order to distinguish contributions to photocurrent from field enhanced electron hole pair generation and carrier injection [3]. X-ray photoemission spectroscopy was used to contrast physicochemical 2DNC-NA bonds vs. polyvinylpyrrolidone coatings representing alternate pathways for carrier injection. A tunable laser scanning confocal microscope (LSCM) was used to measure second harmonic generation of 2DNC-NA hybrids (Fig 3). A computational scaffold supporting analysis spanning atom to nanometer scales was used to develop compact structure-function relations in order to rationalize measured values.

Coordination of EELS, XPS, and LSCM offers a comprehensive approach to characterize increased photocurrent from field enhanced electron hole pair generation and from carrier injection in 2DNC-NA hybrids at high spatiotemporal resolution.

[1] G.T. Forcherio et al., in preparation. [2] J.R. Dunklin et al., in preparation. [3] D.K. Roper et al., in preparation.

497441 High-Efficiency Dynamic Lighting with Transition Metal Elements As Sensitizers

Abstract:

The lighting industry has drastically evolved over the past decade in order to provide low-cost, energy efficient lighting with high color quality. In particular, the development of white light light emitting diodes (LEDs), which employ rare earth (RE) based phosphors to produce a wide range of wavelengths due to their unique optical and electronic properties. Î²-$NaYF_4$:RE nanoparticles (NPs) are currently employed in personal electronics as stable emission centers because of the steady luminescence of RE elements and low phonon matrix energy of Î²-$NaYF_4$ host. Additionally, luminescence can be amplified via co-dopant with a sensitizer agent (Ce^{3+}, Yb^{3+}, and various transition metals (TM)). These RE doped materials are typically coated with a shell layer to passivate the surface sites and limit the external field effects on the RE. In this work, a new class of TM based phosphors is proposed, specifically a-RE doped core with a TM doped shell layer to maintain LED efficiency while simultaneously controlling the allowed luminescent transitions. The ability to control the emission spectra of these RE elements is due to the dynamic nature of the TM absorption with external stimuli. Despite their susceptible nature, there is motivation to use TMs as a replacement for RE sensitizers based on abundance and absorption cross-section.

For this purpose, Î²-$NaYF_4$:Er|TiO_2:Ni core-shell nanoparticles were synthesized by using a combination of thermal decomposition and stÃ¶ber processes. The surface of the shell layer is functionalized with benzoic acid (BZA) ligands via carboxylic acid chemistry in order to create external chemical dipole. Standard structural and compositional characterizations were performed on these nanoparticles in order to identify the morphology, shape, crystal structure and elemental composition. Next, the optical properties, such as absorption and upconversion luminescence, were investigated via optical spectroscopy (UV-Vis/Photoluminescence), demonstrating the ability to shift the absorption spectra of TM ion and tune the RE-TM energy transfer kinetics for controlled upconversion emission. Furthermore, X-ray Absorption Spectroscopy (XAS) was employed to probe the local site chemical environment surrounding the dopant, indicating the ligand induced dipole, for engineering of the dopant oxidation state within the shell layer without changing the overall charge of the system. Moreover, the system is modeled using time-dependent density functional theory (TD-DFT) simulations based on the concepts of Tanabe-Sugano diagrams and Judd-Ofelt theories, showing good agreement between theory and experiment.

500936 Mechanism and Design Principles for Directing Energy Flow in Multicomponent Plasmonic Systems

Abstract:

Plasmonic metal nanoparticles exhibit enhanced light-matter interactions through the excitation of localized surface plasmon resonance. Plasmon excitation results in the confinement of light energy to the nanoparticle surface manifested in high electric field enhancements. This elevated field subsequently decays through either radiative photon scattering or through non-radiative excitation of hot carriers in the bulk material. Many researchers have been attempting to understand the mechanisms of plasmon decay in metallic nanoparticles with the ultimate objective of controlling this decay process.

In this presentation, we shed light on the physical mechanisms underlying plasmon decay in multicomponent plasmonic systems. Using Raman spectroscopy, we show that the introduction of a probe molecule on the surface of plasmonic nanoparticles introduces a new plasmon decay pathway resulting in the selective dissipation of light energy into the molecule. We develop the physical framework governing this energy dissipation and discuss our application of this framework to the design of multimetallic plasmonic nanostructures. We demonstrate that by coating plasmonic nanoparticles with non-plasmonic metals, the plasmon energy is selectively dissipated in the non-plasmonic metal shell. We provide insights into how this physical framework can aid the rational design of multicomponent plasmonic systems for various applications in light harvesting, plasmonics and photocatalysis.

503227 Semiconducting Halide Perovskite Nanomaterials and Heterojunctions

Abstract:

Recently, there is a renaissance of halide perovskite ABX_3 (A = organic ammonium cation, Cs; B = Pb^{2+}; X = Cl^-, Br^-, I^-) as a promising class of semiconductor materials for a variety of optoelectronic device applications. Remarkable features have been discovered in the halide perovskite system, such as long charge carrier lifetime/diffusion length, large absorption coefficient, high photoluminescence quantum efficiency, etc. So far, most of work focuses on bulk crystals and polycrystalline thin films. Achieving new nanoscale morphologies (0D, 1D, or 2D) is of great interest because it offers a unique avenue to depart from the properties of the bulk materials and to study nanomaterials on a fundamental level in order to enable the incorporation of these materials into next-generation devices.

In this presentation, we demonstrate a novel substrate mediated crystallization method to growth high quality one dimensional nanowires with diameters ranging from 200 to 800 nm and lengths over 10 Î¼m, and two dimensional sheets with lateral dimensional up to 20 Î¼m and thickness down to one atomic layer (*Science 2015, PNAS 2016*). Such ionic semiconducting nanomaterials with excellent optical and electronic properties provide a versatile platform for understanding alloy and heterostructure formation. By combining with nanofabrication techniques, we further demonstrate spatially resolved anion exchange chemistry to achieve single crystalline halide perovskite heterojunctions ($CsPbBr_3$-$CsPbCl_3$ and $CsPbBr_3$-$CsPbI_3$) with sharp interfaces and multiple colour emissions (*PNAS 2017*). Such heterojunctions are useful in high performance nanoscale optoelectronics, including highly stable solar cells and colour tunable LEDs/Lasers.

AUTHOR INDEX

AUTHOR INDEX

AIChE
120 Wall Street, FL 23
New York, NY 10005-4020

ISBN 978-1-5108-5783-4
9 781510 857834

CURRAN ASSOCIATES INC.
proceedings.com